MATTERING PRESS

Mattering Press is an academic-led Open Access publisher that operates on a not-for-profit basis as a UK registered charity. It is committed to developing new publishing models that can widen the constituency of academic knowledge and provide authors with significant levels of support and feedback. All books are available to download for free or to purchase as hard copies. More at matteringpress.org.

The Press' work has been supported by: Centre for Invention and Social Process (Goldsmiths, University of London), European Association for the Study of Science and Technology, Hybrid Publishing Lab, infostreams, Institute for Social Futures (Lancaster University), OpenAIRE, Open Humanities Press, and Tetragon, as well as many other institutions and individuals that have supported individual book projects, both financially and in kind.

We are indebted to the ScholarLed community of Open Access, scholar-led publishers for their companionship and extend a special thanks to the Directory of Open Access Books and Project MUSE for cataloguing our titles.

MAKING THIS BOOK

Books contain multitudes. Mattering Press is keen to render more visible the unseen processes that go into the production of books. We would like to thank Joe Deville, who acted as the Press' coordinating editor for this book, the two reviewers Javier Lezaun and a second reviewer who chose to remain anonymous, Jennifer Tomomitsu for the copy editing, proof reading and book production work, Alex Billington and Tetragon for the typesetting, and Will Roscoe, Ed Akerboom, and infostreams for their contributions to the html versions of this book.

COVER

Cover art by Julien McHardy.

ENGINEERING THE CLIMATE

Science, Politics, and Visions of Control

JULIA SCHUBERT

First edition published by Mattering Press, Manchester.

Copyright © Julia Schubert, 2021.
Cover art © Julien McHardy, 2021.

Freely available online at matteringpress.org/engineering-the-climate

This is an open access book, with the text and cover art licensed under Creative Commons By Attribution Non-Commercial Share Alike license. Under this license, authors allow anyone to download, reuse, reprint, modify, distribute, and/or copy their work so long as the material is not used for commercial purposes and the authors and source are cited and resulting derivative works are licensed under the same or similar license. No permission is required from the authors or the publisher. Statutory fair use and other rights are in no way affected by the above.

Read more about the license at creativecommons.org/licenses/by-nc-sa/4.0/

ISBN: 978-1-912729-26-5 (ppk)
ISBN: 978-1-912729-27-2 (pdf)
ISBN: 978-1-912729-28-9 (epub)
ISBN: 978-1-912729-29-6 (html)
DOI: http://doi.org/10.28938/9781912729265

Mattering Press has made every effort to contact copyright holders and will be glad to rectify, in future editions, any errors or omissions brought to our notice.

CONTENTS

List of Figures	6
List of Tables	7
List of Abbreviations	8
Introduction	11

PART I: A 'BAD IDEA' BREAKS INTO POLITICS

1 · Confronting the Crisis	35
2 · The Emerging Politics of Climate Engineering	50
Conclusion to Part I	83

PART II: EARLY VISIONS OF CONTROL

3 · Where Does the Story Begin?	89
4 · Years of Fracture	108
Conclusion to Part II	137

PART III: ENGINEERING THE CLIMATE — SCALING THE ISSUE AND SUGGESTING CONTROL

5 · Assembling an Engineering Problem	143
6 · Devising a Project of Climatological Cultivation and Control	174
Conclusion to Part III	214
Conclusion: Scientific Expertise and the Politics of a 'Bad Idea Whose Time Has Come'	221
Appendix: Document Corpus	239
References	249
Acknowledgments	273

LIST OF FIGURES

FIG. 2.1 Climate Engineering in US Climate Policy (FDsys)   36
FIG. 3.1 Climate Engineering Proposals   58
FIG. 3.2 Virtual Rendering of What Carbon Engineering's Large-scale Direct Air Capture Plants Will Look Like (Credit: Carbon Engineering Ltd.)   62
FIG. 8.1 The Career of Climate Engineering in US Policy   223

LIST OF TABLES

TABLE 3.1 The 'Official Record' on Climate Engineering in US Climate Policy — 52

TABLE 6.1 Expert Witnesses Mentioning Climate Engineering (2006–2009) — 155

TABLE 7.1 Staged Advice (2009–2017) — 190

LIST OF ABBREVIATIONS

ACI	Atmospheric Climate Intervention
ARPA	Advanced Research Projects Agency
AWS	Air Weather Service
BESK	Binary Electronic Sequence Calculator
CCS	Carbon Capture and Storage
CCSP	Climate Change Science Program
CCTP	Climate Change Technology Program
CDR	Carbon Dioxide Removal
CDRMIP	Carbon Dioxide Removal Model Intercomparison Project
CO2	Carbon Dioxide
CRS	Congressional Research Service
DAC	Direct Air Capture
DARPA	Defence Advanced Research Projects Agency
DOD	United States Department of Defence
DOE	United States Department of Energy
DOS	United States Department of State
ENIAC	Electronic Numerical Integrator and Computer
ENMOD	Environmental Modification Convention
EPA	Environmental Protection Agency
FDsys	Federal Digital System
GAO	Government Accountability Office
GCEP	Global Climate and Energy Project
GCM	General Circulation Model
GeoMIP	Geoengineering Model Intercomparison Project
IGY	International Geophysical Year
IPCC	United Nation's Intergovernmental Panel on Climate Change
LLNL	Lawrence Livermore National Laboratory

LIST OF ABBREVIATIONS

NAS	National Academy of Sciences
NASA	National Aeronautics and Space Administration
NASEM	National Academies of Sciences, Engineering, and Medicine
NCAR	National Center for Atmospheric Research
NGO	Non-Governmental Organisation
NIAC	National Institute for Advanced Concepts
NOAA	National Oceanic and Atmospheric Administration
NSF	National Science Foundation
NWS	National Weather Service
OF	Ocean Fertilisation
OSTP	Office of Science and Technology Policy
PSAC	President's Science Advisory Committee
RAND	Corporation Research and Development Corporation
ScoPEx	Stratospheric Controlled Perturbation Experiment
SPICE	Stratospheric Particle Injection for Climate Engineering
SRM	Solar Radiation Management
UK	United Kingdom
UNFCCC	United Nations Framework Convention on Climate Change
US	United States of America
USDA	United States Department of Agriculture
USGCRP	US Global Change Research Program
USGS	United States Geological Survey
WHOI	Woods Hole Oceanographic Institution
WMO	World Meteorological Organization

INTRODUCTION

IN NOVEMBER 2018, NATURE FEATURED SOME SENSATIONAL NEWS: 'FIRST sun-dimming experiment will test a way to cool Earth'. The article talks about a project called Stratospheric Controlled Perturbation Experiment – ScoPEx, for short – a scientific experiment in development at Harvard University. Underneath the headline, the story is illustrated by a photograph. The picture invites us into a Harvard lab to meet ScoPEX's creators. Our view of the team of researchers is partially obstructed by cables and machinery, literally framing the scientists with techno-scientific hardware. The story begins:

> Zhen Dai holds up a small glass tube coated with a white powder: calcium carbonate, a ubiquitous compound used in everything from paper and cement to toothpaste and cake mixes. ... The question ... is whether this innocuous substance could also help humanity to relieve the ultimate case of indigestion: global warming caused by greenhouse-gas pollution. The idea is simple: spray a bunch of particles into the stratosphere, and they will cool the planet by reflecting some of the Sun's rays back into space.[1]

The article introduces us to 'a broader class of planet-cooling schemes ... that have long generated intense debate and, in some cases, fear': so-called *climate engineering* or *geoengineering* measures.[2] The text quickly pushes us down the rabbit hole that lingers beneath this controversial set of labels. On the one hand, the article is about a balloon that would squirt out a substance, usually found in 'everything from paper [...] to cake mixes'. It describes a harmless scientific experiment, exploring an 'innocuous' chemical substance. On the other hand, the text suggests quite the grand mission. It promises the first 'sun-dimming experiment', testing nothing less than a way 'to cool Earth'. It talks about a 'thermostat' to counteract dangerous global warming. It discusses a new remedy against global warming, offering some much-needed relief for the future of humanity by tackling one of the most pressing challenges of our time.

This *Nature* feature thus confronts us with two somewhat conflicting faces of climate engineering – one light-hearted, one grave. These two faces of climate engineering are further illustrated by the self-descriptions of the researchers, as featured later in the article. Dai, on the one hand, is quoted as 'not stress[ing]' about the critique of this research: 'I'm studying a chemical substance [...] It's not like it's a nuclear bomb'. Frank Keutsch, in contrast, is described as 'a reluctant geoengineer'. He engages with climate engineering research as 'he worries about where humanity is heading, and what that means for his children's future'.

If we dig a bit deeper into the world of climate engineering, the picture only gets more complicated. Almost a decade prior to the publication of the *Nature* piece, in October 2010, the public was similarly prepared for 'the first field test of a geoengineering technology' – this time in the UK.[3] Back then, it was the Stratospheric Particle Injection for Climate Engineering project – SPICE, for short – which had sought to 'move solar geoengineering out of the lab'.[4] The project proposed a small field study to test the technical feasibility of a balloon deployment system which was initially scheduled for 2012 in Norfolk, England. In this case, too, two seemingly contradictory perspectives clash in an effort to make sense of the planned experiment: In one version, SPICE would have been the first climate engineering outdoor experiment; in another, it would have entailed pumping 'some water – no more than it would take to fill a child's paddling pool' through a kilometre-long hose and produce a cloud of fine mist.[5] Due to public protest and internal issues, the experiment eventually had to be cancelled.

Another two years earlier, in 2008, meteorologist Yuri Izrael and colleagues conducted a climate engineering experiment just outside Moscow. The scientists installed generators aboard helicopters to spray sulphuric aerosol into the troposphere. The team then measured 'basic meteorological variables in the surface atmosphere (stratification, temperature, air humidity, and wind speed) and microphysical and optical characteristics of aerosol particles'. Izrael and his team concluded that they had 'shown how it is principally possible to control solar radiation passing through artificially created aerosol formations in the atmosphere with different optical thickness'.[6] The scientists published their findings in two papers. While the publication that reported the first part of the experiment did not connect the project explicitly to solar geoengineering, the one that reported the second part of the experiments did do so.[7]

Now, if we continue our search a bit further, the picture gets even more complicated. Not only do we find that there have been a bunch of field studies, experiments, and demonstration facilities that are directly relevant to climate engineering research – some of which date all the way back to the 1990s. But we find that there is, in fact for years now, a variety of research projects being conducted on the potential to deliberately intervene in and control the Earth's climate. These projects range from modelling studies and computer simulations to social scientific, economic, and ethical research. Part of this research attracted little in the way of controversy as it was simply conducted under different labels; it was not called climate engineering. And part of this research was deemed unproblematic as it was theoretical and conducted 'inside'.[8] Phil Willis, Chairman of the UK Science Committee illustrated this latter point in 2009 when he stated that the British government 'wholeheartedly supports' any 'research that uses computers to model the impact of geoengineering technologies'. That research, with '*a real impact on the wider climate*', however, should be subjected to international regulation.[9] This implicit heuristic of climate engineering research raises the question of exactly what ontological line is crossed between lab and field, between 'indoor' and 'outdoor'? How is 'curiosity-driven indoor research' unproblematic, irrespective of its particular intent or purpose, and yet, 'outdoor research' is deemed 'unethical' without the necessary regulation?[10] Where does climate engineering move from theory to 'real world'? How can one even conceive of the 'real impact' of research?

These questions and examples begin to show just how difficult it is to precisely grasp what climate engineering *is*. The current status of climate engineering, particularly as it is portrayed to the public, seems to oscillate between two extremes. On the one hand, climate engineering is presented as 'just' research: it is not really existent yet, and therefore seems harmless. On the other hand, it is envisioned in its future reality as a 'global thermostat', a serious and almost miraculous tool, bringing about grave societal consequences and therefore raising fundamental normative questions.[11]

These two extremes seem to neatly separate the science of climate engineering from its politics, giving each its own place in time: *today*, climate engineering extends to nothing more than harmless science, but *in the future*, it will fundamentally change the politics of climate change. Narratives that use these extremes separate science and politics, only to connect the two in a linear fashion: scientific research supposedly creates the base for political decisions that might then be taken at some point in the future. The director of a national research laboratory in the United States which was recently awarded federal funds to explore the merit of climate engineering told the *Scientific American*, 'one of the things I'm interested in doing is let's separate the science out'. The director engages here in what Stephen Hilgartner (referring to Bruno Latour) has described as a key role of science advice. That is, he seems to be seeking to 'purify' the issue of climate engineering from its political components.[12] Following this line of reasoning, the director emphasised that the support for this research should not be mistaken with policy-level approval of such measures. According to the magazine, the goal is rather 'to give policymakers a clear view of how a hurry-up bid to save the planet would work'.[13]

Typically, the questions that dominate public debate around climate engineering, therefore, go something like this: will it actually be possible? What is the research that we will need to be able to answer this question? How could we 'test' this technology without actually deploying it? Would such a thermostat be ethical? Who would control this thermostat? Which regulatory frameworks are in place to govern it and with which consequences?

With this book, I want to suggest a change in perspective. While questions like these are obviously important, they run the risk of steering attention away from what is at stake, which is the very essence of the proposed technology. To use the often-invoked metaphor of the global thermostat, what is at stake here is not merely the shape and form, but the very existence, the making and assembling, of this thermostat. Instead of debating the implications and consequences of this techno-political project, this book unpacks its genesis and seeks to understand how we got here in the first place. This book traces how we got to make sense of and problematise climate change in this particular manner; it unpacks how this vision to engineer the climate has and continues to unfold its saliency and meaning.[14]

Climate engineering is not a sun shield, ready to be wielded by politicians. Most of the approaches being discussed do not exist as ready-to-be-deployed or even ready-to-be-tested tools, as technologies merely awaiting the push of a button. Notions such as 'global thermostats' or 'sun dimming experiments' are problematic in this sense and have been rightfully criticised in recent years. Yet, the alternative suggestion – that climate engineering is 'just research' – equally misses the point.

Since at least 2009, climate engineering has incrementally arrived on political agendas at local, national, and transnational levels.[15] State and non-state actors around the globe have begun to consider such measures in various ways. Private investors and energy companies, too, have pushed the development of approaches to effectively remove carbon dioxide from the atmosphere and are seeking ways to commercialise such technologies. In a widely-publicised move, Elon Musk, for example, announced 'the largest incentive prize in history' in January 2021 – the XPRIZE Carbon Removal initiative. With this initiative, Musk is promising $100 million for whomever can come up with 'real systems' that are able to permanently remove CO_2 from the atmosphere. XPRIZE describes its efforts as seeking to 'tackl[e] the biggest threat facing humanity – fighting climate change [...]'.[16]

Aside from efforts to commercialise and develop climate engineering technologies, we can observe how these approaches are beginning to become institutionalised as standard means to counteract global warming at the international level. Since 2018, the United States, for example, initiated efforts to integrate 'climate coolants' into the International Standardization Organization (ISO) as a way to offset greenhouse gas emissions. This was a highly controversial effort to normalise solar climate engineering measures in order to mitigate global warming. In addition, the United Nation's Intergovernmental Panel on Climate

Change (IPCC) increasingly incorporates climate engineering approaches into its climate change scenarios.[17]

Finally, and maybe most significantly, governments around the world have begun to establish research and development programs for a number of years now. The United Kingdom, the United States, and Germany have been most active in this context with government authorities conducting official inquiries into the issue of climate engineering since around 2009. As we will see in more detail in Chapter 2, the United States initiated its official exploration of the controversial topic in 2009 together with the UK, and shared witnesses and insights about their respective assessments. Germany followed suit when representatives brought the issue up during parliamentary debates in 2011 and 2012. Since then, climate engineering has spread into policy agendas around the globe.

In 2013, the Russian government attracted controversial media attention when leaked documents suggested that it had urged the IPCC to include the controversial measures into the organisation's much-anticipated fifth assessment report on the status of climate change science and expertise. Since 2015, both the Indian and the Chinese government have developed climate engineering programs. Examining both the physical mechanisms and the governance implications of these measures, the Chinese Ministry of Science and Technology initiated what has been characterised as one of the world's largest state-funded climate engineering research programs at Beijing Normal University. And in March 2020, Australia conducted a cloud-brightening experiment off Queensland to test if ocean cooling mechanisms could serve to save the Great Barrier Reef.[18]

In the face of these developments, the question is how might we gain a better and more meaningful understanding of climate engineering, one that goes beyond false dichotomies of science versus politics, theory versus real world, indoor versus outdoor research, and the choicelessness of prescribed futures? As the examples introduced so far begin to show, climate engineering is being developed across a large geographical scale, although not necessarily in oceans and atmospheres (yet), but rather in geographically distant spaces of political decision making. The often-invoked binaries precisely seem to miss this point.

It is not only 'outdoor experiments' that are able to *'change the world'*.[19] Beyond 'just research', beyond particle flows, beyond the mere 'dispersion of material in the environment',[20] climate engineering is gaining political traction and has concrete political impacts. It is beginning to take concrete shape; it is being assembled and installed, fine-tuned and mounted as a potential policy measure, and as a controversial tool for counteracting anthropogenic climate change.

To properly grasp and engage with this reality of climate engineering, we need to understand how it came to be. In other words, to decide if and how to move forward, we need to understand how we got here. Instead of essentialising climate engineering as either 'just research' or a miraculous tool, we need to take a look back. We need to understand how climate engineering became what it is today, how it became assembled in its current form, how it became established as a potential policy tool, and how it became a controversial response measure to the issue of global warming. This book seeks to contribute to current debates over climate engineering by unpacking its genesis. It is about the 'career' of climate engineering in the United States, from curious scientific idea to serious politics.

RE-CONTEXTUALISING THE RISE OF CLIMATE ENGINEERING: THE ARGUMENT IN A NUTSHELL

The current debate over climate engineering generally concerns two broader sets of technological concepts. On the one hand, so-called carbon dioxide removal (CDR) approaches seek to counteract climate change by sucking CO_2 from the atmosphere. On the other, so-called solar radiation management (SRM) measures aim to counteract climate change by reflecting incoming sunlight back to space.[21] We will come back to the technical intricacies and definitional struggles over these approaches in Chapter 2. Exactly how these efforts are categorised and labelled are precisely part of the story of this book. For now, I simply want to emphasise that the umbrella terms of climate- or geoengineering bundle together a wide variety of research contexts in their differently articulated promise to fundamentally alter the politics of climate change. These labels are not primarily 'scientific'; they do not match disciplinary boundaries or follow methodological approaches. In fact, they have been criticised in this respect on many fronts.[22] Instead, these labels become meaningful and are subject to heated debate as they bind various lines of scientific inquiry to the societal challenge of tackling climate change. The definitional struggles over what climate engineering is, can, or should be, calibrate this relation of scientific inquiry and political intervention, with each emphasising different sides or aspects of this charged relation. Umbrella terms such as climate engineering or geoengineering might not make a lot of sense from a technical point of view, but their controversial status precisely shows what is at stake here. These definitional struggles hint to a relationship between science and politics that is much more complex than what idealist models of advisor and decision-maker would suggest. In the emerging debate over climate engineering, science does not merely figure as a neutral evidence base for politics; it does not simply prepare a difficult political decision with positive facts. Instead, science and politics are coupled 'upstream'. That is, they are linked in the very formulation of climate engineering as a potential measure to counteract climate change.

So, returning to my point from before, the important question then becomes, how did we get here? How did climate engineering earn a spot in climate policy agendas despite enormous scientific complexities and fierce political contestation? And, more generally, what can we learn from this case about the relationship between science and politics in modern society?

This book explains the conflicted status of climate engineering today with historically grown alliances between climate science and politics. It describes the emergence of climate engineering as a story of the political cultivation of climate science for the state. Over the course of the following chapters, I will unpack the historical genesis and 'career' of climate engineering as a controversial policy tool along two dimensions. On the one hand, we will see that the career of climate engineering evolves along historically particular settings of problematising and making sense of climatic change. It corresponds, in other words, to the contingent history of climatic change as a societal problem. On the other hand, we will see how these historically particular settings of problematising climatic change directly relate to shifting alliances between climate science and the state. That means that the career of climate engineering unfolds along respectively changing roles of climate expertise within the state.

This perspective thus re-contextualises the history of climate engineering within the bigger history of political efforts to cultivate climate science for the state. It suggests that what we today discuss as climate engineering has historically served as a continuous, yet shape-shifting node, effectively linking scientific to political agendas. In so doing, this perspective emphasises just how interrelated efforts to understand and efforts to govern, even control, the climate have been.

I approach this account from two angles by connecting insights from historical scholarship with perspectives from science studies and historical sociology. On the one hand, I draw from historical analyses to demonstrate that the recent rise of climate engineering as a controversial last resort measure or 'Plan B' against dangerous climate change is only the latest chapter in a much longer standing career of efforts to intervene in, modify, and even control the Earth's climate. This career bundles the disparate histories of various technological concepts and scientific insights, sprouting in different contexts and times, under different labels, serving different political purposes. On the other hand, I zoom into the recent debate over climate engineering in US politics. I trace how these measures arrived on the US political agenda around 2009 as a potential approach against anthropogenic climate change. By following these debates through the concrete arenas of climate policy making, I seek to unpack how science and politics precisely connect to one another in the making of this controversial policy tool. With this analysis, the book suggests that what is at stake in this debate over climate engineering is just as much a political vision as a scientific project. Looming large in this debate are thus different visions for the role of science in addressing one of the most pressing challenges of our time. Unpacking the historically grown role of science in both defining and tackling this issue is essential for enabling a meaningful debate over what this role should look like in the future. There was, and there will be, no point zero at which decision-makers will take an informed decision on how to proceed. As much as politics might sometimes allude to external urgencies that force our hands, climate engineering was not infused into the political process by the external urgency of dangerous climate change. It arrived here from within: this particular vision of making sense of and responding to climatic change has a historical legacy and system. It is the result of established and cultivated science-politics alliances.

SCIENCE, POLITICS, AND THE CAREER OF CLIMATE ENGINEERING: THE ANALYTICAL FRAMEWORK

In the following, I briefly introduce the central concepts that guide this book's analysis. I explain what I mean by the notion of the 'career' of climate engineering, what this conceptual take implies for the study of scientific expertise in politics, and what data the following analysis rests on.

Dissecting the 'career' of climate engineering

This book unpacks how we got here. It seeks to understand how climate engineering arrived on mainstream climate policy agendas and how it became established as a controversial policy tool against global warming. To do so, the book follows climate engineering on its turbulent journey at the interface of science and politics; it traces, in short, what I call the 'career' of climate engineering.

This notion of the 'career' of climate engineering, first of all, marks a particular choice of perspective. It suggests a distinct methodological approach to climate engineering as the object of this study. Deciding what climate engineering *is*, is far from straightforward as we have seen at the outset of this introduction. Writing a book on its career in this sense means to work with this ambiguity.

This book chooses as its starting point a set of distinct contexts and arenas in which climate engineering has taken concrete shape as a potential policy tool against global warming to then unpack how it came to be assembled precisely as such: what kinds of experts and actors were relevant in devising this policy tool? What kinds of expertise, what kinds of global political and historical contexts, and what kinds of observational devices and policy programs were at play here? This perspective gives climate engineering a life of its own, so to speak. Connecting to Gil Eyal's call, *For a Sociology of Expertise*, retracing this career of climate engineering entails 'a history without a protagonist'[23] in the sense that it is not primarily the story of a group of individuals or experts, but instead one that places the historical emergence and trajectory of these measures as policy tools front and centre. Instead of asking how a certain group of experts managed to push climate engineering into the political limelight, this book rather asks how the suggestion of climate engineering managed to push a certain group of experts into the political limelight. The idea, in short, is to give this notion of climate engineering a kind of historical agency for itself and to ask how this notion in its current shape and form has gained political traction, how it has bound scientific to political agendas, and how it has made certain people into experts and certain modes of observing the world into expertise.

The theoretical point, then, is not to contrast real and objective issues with socially constructed ones. I do not wish to show that the issue of climate change or the suggested response of climate engineering is 'constructed' at its core, whatever that might mean.[24] Instead, the point is to qualify *how* climate change came to be seen and understood as an issue that would lend itself to deliberate climatological intervention and control. Following a point made by Bentley Allen, the point is to understand how climate change became 'assembled'[25] in the political realm as an issue of techno-scientific intervention and control and how this particular gaze onto the issue has defined shifting alliances between climate science and the state.

This perspective on the career of climate engineering then secondly implies a particular approach to studying the interrelation between science and politics. The analysis in this book connects scholarship on the emergence of societal issues and the making of governance objects with insights from the sociology of expertise, science studies, and science and technology studies (STS).[26] The book traces the career of climate engineering by asking how these measures have historically linked scientific and political agendas. The career of climate engineering thus serves as a prism for the diverse and historically particular alliances between science and politics that have eventually brought forth this controversial response measure. As we will see throughout the following chapters, the dynamic trajectory of this response measure challenges linear conceptualisations of the science-politics nexus and instead emphasises reciprocity. This trajectory is neither shaped primarily by political will – for example, because politicians decide a certain issue or response measure is, or is not, of particular relevance – nor does it strictly follow the scientific 'discovery' of new puzzles and problems. Instead, the approach developed over the course of this book demonstrates that the recent rise of climate engineering has linked science and politics *reflexively*; it shows how the formulation of both problem and response has emerged from the mutual interdependence of both societal spheres. The shifting historical contexts of assembling climate change as an engineering challenge thus serve to illustrate just how deeply intertwined efforts to understand and efforts to govern the climate have been.

Reflexivity and interdependence, however, should not be mistaken with the dissolution of differences in this context. I do not wish to draw attention to fuzzy boundaries or suggest that science and politics have somehow become the same. By drawing on differentiation-theoretical accounts that have been developed within sociology, my goal is rather to gain a better understanding of how science and politics as two societal spheres interrelate with one another.[27] That means understanding how science and politics latch onto one another in devising this controversial policy tool and how this controversial policy tool matches distinctly scientific to distinctly political struggles. Examining the interdependence of science and politics in this sense importantly contributes to a better understanding of what drives science and politics as distinct spheres of societal communication.

Scientific expertise

Tracing the historical trajectory and career of climate engineering thus requires tracing historically specific modes of problematising climatic change. And this entails looking at distinct groups of people, experts, and institutions as much as looking at particular modes of observing and studying, of knowing and governing climate change. Two concepts therefore guide my analysis of the role of scientific expertise in shaping this career of climate engineering: the concept of the *expert infrastructure* and the concept of *expert modes of observation*.

If we understand science as a system or network of communications, the question of scientific expertise in politics, first and foremost, becomes a relational one: how does scientific expertise become politically relevant? Or, to put it the other way around, where does policy-relevant scientific expertise come from?[28] Who are relevant experts and who decides? The concept of the *expert infrastructure* seeks to capture and build on this relational dimension of scientific expertise. It seeks to draw our attention to the structures that link scientific expertise to politics, to the institutionalised settings, the advisory panels, the expert organisations, and assessment procedures that effectively bring scientific expertise to bear on politics – whether intended or not. The concept thus seeks to shed light on the empirical diversity of the formal and informal structures, the programs and agencies, and the 'cliques' and networks that have put climate engineering on the political agenda. We may picture a kind of transport infrastructure in this context, an infrastructure that is made up of routes and paths, of bike lanes and motorways, train tracks and shipping routes that each display different modes of transportation yet are all somehow connected.

The concept of *expert modes of observation* marks the epistemological dimension of scientific expertise. This concept understands scientific expertise as a distinctly formalised mode of observing the world, as a particular form of structured observation.[29] It asks empirically for the various ways in which expert observations on engineering the Earth's climate are formalised or structured, and how, in turn, these modes of observation shape the politics of climate engineering. The concept thus seeks to bring into focus questions about the social, historical, and even material preconditions of such expert modes of observing. Speaking with Paul Edwards, we might refer to a 'vast machine' to describe the complex of computers, satellites, measuring devices, theories, models, experiments, threshold values and tipping points that formulated the challenge of governing climate change as a challenge of engineering the Earth's climate.[30]

This reflexive connection of science and politics is illustrated, for example, in the analogous titles given to two accounts on the subject: Jim Fleming's *Fixing the Sky* and Timothy Mitchell's *Fixing the Economy*. Both accounts carve out how different forms of scientific observation – atmospheric and economic – not only provide academic insights, but also transform their respective subjects politically. As a new form of observing, both atmospheric and economic expertise generate new territories of governance and control.[31] Scientific modes of observation in this sense become politically relevant as they make issues politically legible; that is, as they render issues governable and suggest control. Mitchell holds that economics provided 'a new language, in which the nation-state could speak for itself and imagine its existence as something natural [and] subject to political management'.[32] This formation of economics as a scientific field, in turn, critically depended on the discovery of the economy as a political subject, according to Mitchell. In this sense, the concept of the modes of observations draws attention to the fact that scientific expertise shapes the politics of societal challenges beyond the neat settings of advisory processes or expert commissions. It emphasises the interdependence between observing and addressing societal challenges, such as climate change.

The distinction between expert infrastructure and expert modes of observation thus follows a similar rationale as the distinction between scientific 'expertise' and scientific 'expert', as called for by some authors within the sociology of expertise.[33] It assumes that the relationship between scientific expert and scientific expertise is not unidimensional but runs both ways. It is not only the expert who defines what counts as expertise, but it is also expertise that defines who counts as an expert. Taken together, the two concepts shed light not only on the prominent channels and the hidden byways that bring scientific expertise to bear on politics. They also illustrate the particular distinctions, the methods, theories, or instruments that have cast the challenge of tackling climate change in the terms of scientific intervention and control.

EMPIRICAL APPROACH AND MATERIAL

There are, of course, many ways to approach and study this career of climate engineering as suggested by the theoretical categories and dimensions described above. In this book, I choose a national policy context as the starting of my analysis, namely the policy context of the United States of America. The United States provides a dynamic breeding ground for the exploration of climate engineering as a climate policy tool, and it is therefore an interesting context for studying these measures. For one, climate change remains a partisan issue in the United States. Despite this fact, the United States has played an essential role within global efforts to tackle this issue (for good or ill). And two, the scientific community researching climate engineering is comparatively active here, while the political debate on these measures remains highly contested.

This choice of perspective means that we will begin our inquiry in Part I of this book by asking what the current status of climate engineering is in this particular policy context. In other words, what do 'the politics' of climate engineering look like and what do they amount to? What are the kinds of political decisions that are currently made, the policies that are discussed, the reports that are written up, and the hearings that are being held on climate engineering? What, in other words, is the essence of these emerging politics of climate engineering?

Starting from these kinds of questions, we can then set out to examine throughout the rest of this book how we got here. To do so, this book draws on a variety of empirical material. Most centrally, this includes a corpus of policy documents that spans 30 years, which I will describe in more detail shortly (for an overview of these documents, see Appendix). For the time before the 1990s, I complemented the observations from this document corpus with an analysis of politically commissioned scientific assessments and historical analyses. Furthermore, the book rests on insights from a research stay at the Cooperative Institute for Research in Environmental Sciences (CIRES) at the University of Boulder, Colorado, as well as participatory observations during three climate engineering conferences and 15 unstructured expert interviews.

The document corpus provides the primary database for my analysis of how climate engineering took shape, and how it unfolded its saliency and meaning, in US climate policy (see Chapters 1, 2, 5, 6). It comprises all federal proceedings, dealing with the issue of climate engineering between 1990 and 2020, as documented in the Federal Digital System (FDsys). FDsys is a public digital archive that preserves official records from the federal government of the United States. It maintains digital access to a vast spectrum of policy documents, from congressional records, appropriations, hearings, to suggested bills, rules, or entries to the Federal Register.[34] With this document corpus, I sought to construct a kind of 'window' onto the US policy process – a window that would allow me to examine the particular arenas in which climate engineering took shape in the US political context over the years and also empirically determine the relevant expert infrastructure and the defining expert modes of observation that defined this process.[35] Let me illustrate what I mean by that.

To begin with, these policy documents bundle the many layers of US federal policy making. This means that they point us not only to the main policy bodies that are exploring and processing climate engineering in some way or another, like the congressional commissions which hold hearings or decide on funds, or the Representatives and Senators who introduce legislation, or the Executive agencies that propose rules. But these documents also track the many different state and non-state actors, the global governance bodies, frameworks, and reports that inform and guide US policy from 'outside' the federal policy process. In her instructive study on *The Politics of Objective Advice*, Ann Keller, for example, illustrates how congressional hearings point to 'scientific findings, shifting analytical uncertainties, emerging regulatory approaches, […] international negotiations, evolving political positions and arguments' as they all somehow shape legislative politics.[36] Despite choosing a national policy context as the starting point of the analysis, this material also sheds light on the role and relevance of the international and global dimension of climate engineering. The policy documents suggest how the career of climate engineering evolves between the imperatives of national decision-making on the one hand, and the multi-faceted institutions of global climate politics, on the other. We will see how crucial a global understanding of climatic change is to this vision of technoscientific climate intervention, but also simultaneously how this vision caters to political hopes of national control in the face of this global challenge – how it takes shape, in other words, as a tool for the state.

The emerging politics of climate engineering become visible here as a network of communication. This network comprises all kinds of communicative processes that provide the societal capacity of taking collectively binding decisions, including the individuals or organisations that provide these communications, and the policies, regulations, frameworks, and so on, that stabilise them.[37] The essential role of science in these emerging politics of climate engineering is observed through this lens of the policy-process. It manifests in the experts and modes of observations that effectively shape this policy process.

Furthermore, these policy documents mark the temporality of the career of climate engineering in US policy.[38] Hearings, rules, entries to the federal register, and so forth document policy communications in relation to a particular point in time. It seems to lie in the very essence of files, and especially the public record, to 'capture' a particular moment. Policy documents thus help us chart the temporal trajectory of this career of climate engineering all the while mapping the many actors and organisations that have stabilised this particular trajectory.

And finally, these documents comprise official and public communication. Congressional hearings, for example, are directed at the general public, a particular committee, an agency, Congress, or the White House. They are held and documented 'to communicate something publicly'. What we see through this window is a 'purposive arena' of political communication.[39] It allows no glance behind closed doors; it does not enable peeking into the minds of policymakers, but it documents how climate engineering has been established in the political arena over the years. The corpus traces shifting discursive frames and arenas of contestation. Just like the status of scientific publications for studying science, the corpus is instructive not despite, but precisely because of its orchestrated nature.[40]

STRUCTURE OF ARGUMENT

The three main parts of this book will guide us through the career of climate engineering by dissecting some of its defining historical settings between the turn of twentieth century and the first decades of the new millennium. Over the course of the following chapters, we will see how each of these historical settings or 'stages' in the career of climate engineering corresponds to a historically particular mode of problematising climatic change as well as shifting alliances between climate science and the state.

We begin this inquiry in November of 2009, in *Part I* of this book. This is our starting point; this is when climate engineering officially arrived in US climate policy as an issue in its own right and assumed the form and status that still defines the debate over these measures today. Chapters 1 and 2 in this sense set the stage and present the basic premise of this book. In Chapter 1, the book contextualises the sudden rise in US political attention to climate engineering around 2009. We will see how climate engineering took shape and gained political traction at this point in time as a 'bad idea whose time has come'.

Chapter 2 complements the general outlook on this historical setting of the career of climate engineering by zooming into the concrete arenas of US policy making in which these measures began taking political shape during these years. We will see how, on the one hand, climate engineering materialised as a matter of fact(s) with policymakers and experts establishing an 'official record' on the issue, engaging them in definitional struggles over what climate engineering is, should, or can be. On the other hand, climate engineering became structurally internalised into the federal infrastructure. We will see how it took shape as a set of techno-scientific challenges that began to guide political efforts at cultivating relevant climate engineering expert capacities.

In *Part II* of this book, we will travel back in time to try to determine the historical roots of this 'bad idea whose time has come'. This move makes it possible for the book to show just how deeply intertwined efforts to understand and efforts to govern climatic change have always been. Chapter 3 sends us on a sweeping journey all the way back to the turn of the twentieth century, when human impacts on the climate were beginning to become systematically explored. We will see how before these initial findings on human impacts on the climate were problematised, they provoked positive techno-scientific visions of targeted modification and control. Climatic change, in other words, appeared as a potentially grand opportunity for humankind during these years. The geopolitical challenges of the first half of the twentieth century would only bring this dynamic into full swing, and, as a result, begin to establish climate science as a critical tool for the state, promising the deliberate modification, even control of climatic conditions for military and national strategic purposes. Chapter 4 traces a fundamental shift to this setup, both regarding this problematisation of climatic change and, correspondingly, in the defining alliances between climate science and the state. The chapter illustrates how the politicisation of global warming as an environmental issue during the 1970s and 1980s drowned out the previous techno-optimism, and with it, political excitement over the potential of deliberate climate modification and control. What we discuss today as climate engineering, in other words, did not gain, but rather lost currency in the face of political concerns over dangerous climate change. The response did not quite fit the problem (yet). We will see how climate science no longer seemed to promise techno-scientific control for the state, but instead appeared to question the political and economic status quo. The politicisation of global warming, however, not only fractured established alliances between climate science and the state, but it also forged new ones. The chapter suggests how in this particular historical setting, climate science became established as the problem-defining authority for this newly politicised issue.

Part III of the book zooms once more to the exploration of climate engineering in US climate policy, continuing to trace the career of climate engineering from the turn of the new millennium to its first decades, returning to complete the story that began in Part I. We will see how climate engineering re-gained political traction during the early 2000s when the very problem that these measures promised to address was reformulated. Chapter 5 illustrates how climate engineering moved further into the political limelight when climate change became assembled as a challenge that would lend itself to technological intervention and control. Climate science shifted its status in this context from problem-defining to problem-addressing authority. Beginning in 2009, climate engineering then fully arrived on the US political agenda as an issue in its own right.

In Chapter 6, the book comes full circle. Building on the observations from Chapter 2, we will delve further into the role of scientific expertise in the politics of climate engineering, isolating the particular modes of expert observation, as well as the defining expert infrastructure that undergirded this most recent stage in the career of climate engineering. Chapter 6 suggests how the recent rise of climate engineering provides a kind of synthesis that reconciles two historically conflicting roles of climate science within the state. In its outlook as a 'bad idea whose time has come', climate engineering aligns the initial hopes of techno-scientific control over the climate that have shaped political interest in climate modification for the first half of the twentieth century, with the critical positions of climate scientists and environmental movements, emphasising the limits of techno-scientific control during the second half of the century.

The Conclusion provides some reflections on the book's analysis. With the book being finished in the midst of the Covid-19 global pandemic, climate engineering seems to fit eerily well into a world that has turned to scientific expertise as a tool of crisis aversion. I suggest that this perspective on the career of climate engineering not only sheds light on a highly controversial and somewhat curious debate within climate policy, but that it critically speaks to the status and role of scientific expertise in contemporary politics more broadly.

NOTES

1 Tollefson (2018). For an overview of the ScoPEx project, see, e.g., Dykema and others (2014).
2 I will use the term 'climate engineering' throughout this book as I find that it captures most accurately what is at stake in these debates – namely efforts to deliberately intervene in and potentially control the Earth's climate.
3 Stilgoe (2015).
4 SPICE was a collaboration of the University of Bristol, the University of Cambridge, the University of Oxford, and the University of Edinburgh (see SPICE (2018); see also Hulme (2014: 57) or Stilgoe (2015) for a detailed overview of the planned experiment; see also Doughty (2019)).
5 Stilgoe (2015: 12).
6 Izrael et al. (2009a: 226, 272). It seems surprising how little attention this experiment has received, especially since Yuri Izrael was a renowned scientist, having served as Vice-Chairman to both the World Meteorological Organization (WMO), as well as the IPCC (World Meteorological Organization (2019); see also Doughty (2019: 102)).
7 Izrael et al. (2009b)
8 See, for example, the EPEACE project, which generated relevant insights to SRM research (Russell et al. (2013), but also Russell, (2012)). For the case of ocean fertilisation studies, see Lawrence and Crutzen in Blackstock and Low (2019: 90); Williamson et al. (2012). In addition, *Oceanos*, a marine research organisation lists an overview of ocean seeding experiments on their website (Oceanos 2018).
9 Wills in US House of Representatives (2009: 231), emphasis added.
10 Robock and Kravitz in Blackstock and Low (2019: 97, 98).
11 Holly Jean Buck also speaks of 'binaries' in this context (Buck 2019).
12 Stephen Hilgartner (2000: 4).
13 Fialka (2020).
14 See also Brian Wynne in this context, who stresses the importance to study 'the ultimate contingency of saliency and meaning' for science studies (Wynne 2003: 404).

15 An interactive world map which tracks climate engineering projects around the globe is maintained by the Heinrich-Böll-Stiftung, a German think tank affiliated with the Greens' Party, together with the ETC Group, a biotechnology watchdog.
16 XPrize (2021).
17 For the ISO case, see, e.g., International Organization for Standardization 2021; Möller 2021. For a critical account of the IPCC case and its consequences, see, e.g., Beck and Mahony (2018).
18 For the case of the United States, see, e.g., US House of Representatives, Committee on Science and Technology (2009); US House of Representatives, Committee on Science and Technology (2010b). For the case of the UK, see, in particular, Science and Technology Committee (2010). For the case of Germany, see, e.g., Deutscher Bundestag (2010); Umwelt Bundesamt (2011); Die Deutsche Bundesregierung (2012). For an account of the Russian case, see, e.g., Lukacs, Goldenberg, Vaughan (2013); Intergovernmental Panel on Climate Change (2013). For an account of Indian initiatives, see Bala and Gupta (2017, 2019). For the case of China, see, e.g., Edney and Symons (2014); Cao, Gao, and Zhao (2015); Temple (2017); see also Bala and Gupta (2019: 24). Moore et al. suggest that many Chinese researchers were introduced to climate engineering measures through a number of scientific meetings that the Solar Radiation Management Governance Initiative (SRMGI) held in 2011 (Moore et al. 2016). For the Australian experiment, see, e.g., Readfearn (2020a, 2020b). For a comparative overview over the debate of climate engineering in international policy contexts, see, e.g., Huttunen, Skytén, and Hildén, (2015).
19 US Senate (2015: 12), emphasis added.
20 Shepherd in US House of Representatives, Committee on Science and Technology (2009: 110).
21 For a general classification of climate engineering approaches, see, e.g., Royal Society (2009: 6); National Research Council (2015a; 2015b).
22 See, e.g., US National Research Council (2015b: vii). Jim Fleming, for example, points out that '[...] an engineering practice defined by its scale (geo) need not be constrained by its stated purpose [...]' and 'to constrain the essence of something that does not exist by its stated purpose, techniques, or goals is misleading at best' (Fleming 2010: 228).
23 Eyal (2013: 863).
24 This assumption separates this analysis from the agenda of the social problems literature, which has, for a large part, sought to demonstrate the constructivist core of social problems. For an illuminating critique of this strand of literature, see, e.g., Woolgar and Pawluch 1985. Eyal has instructively criticised the false dichotomy between 'what is real/objective and what is merely attributed/socially constructed' in some of these works (Eyal 2013: 864, fn.2).

25 Allan (2017: 131).
26 For the notion of 'assembling' governance objects, see, e.g., Allan (2017). For scholarship that has developed a concept of expertise in relation to societal problems – that is, asking how societal problems become the objects of expert labour – see, e.g., Mitchell (2002); Eyal (2013; 2019); Abbott (2014).
27 For accounts of the interrelation of science and politics as two distinct social systems, see, especially, Weingart (1983; 2001); Luhmann (1990; 2013); Stichweh (2006; 2015). For accounts on the interrelation of science and politics as two distinct fields of social practice, see particularly, Bourdieu (1998; 2004); Baker (2017).
28 For a relational perspective on scientific expertise, see particularly Eyal (2013; 2019); Grundmann (2017).
29 For this notion of structured observation, see, e.g., Luhmann (1990: 645). The concept of modes of observation also relates to Allan's 'modes of abstraction' or Latour's notion of transcriptions. It is about making an issue legible across a variety of contexts (Allan 2017: 138).
30 Edwards (2010).
31 Mitchell (1998); Fleming (2010). See for this context also Scott (1998).
32 Mitchell (1998: 90).
33 See especially Eyal (2013).
34 See Government Publishing Office (2018) for an overview of all available collections. The document corpus comprises 106 documents (see Appendix: Document Corpus for a detailed list of the included records). This book's analysis places a particular focus on the documents before 2015 as it seeks to explain the controversial arrival of these measures as a potential tool against climate change on the US political agenda.
35 Ann Keller fittingly uses the notion of a 'window' in the context of congressional hearings, which provide insights 'into how events both internal and external to Congress shape legislative debates' (2009: 95).
36 Keller (2009: 95).
37 For this concept of politics, see, e.g., the early political sociology of Luhmann e.g. (2015: 35–44); but also, his later monograph on the topic, (2002: 81–88).
38 For the temporality of hearings, see also Keller (2009: 95).
39 Keller (2009: 95).
40 See also Hilgartner (2000), who examined scientific assessment reports to study expert advice as 'public drama'.

PART I

A 'BAD IDEA' BREAKS INTO POLITICS

I
CONFRONTING THE CRISIS

IN NOVEMBER 2009, BART GORDON, THEN CHAIRMAN TO THE US HOUSE Science Committee, welcomed his fellow members of Congress to a session of hearings: 'good morning [...] Today we begin what I believe will be a long conversation'.[1] The 'long conversation' that Gordon initiated here concerned 'the deliberate large-scale modification of the Earth's climate systems for the purposes of counteracting climate change',[2] also referred to as *climate engineering*.

This is where our inquiry begins. As I will suggest in this and the following chapter, Gordon's hearing was a watershed moment for the career of climate engineering in the United States. The hearing marked the arrival of climate engineering on the agenda of US climate politics. Over the following months and years, US policymakers began to assess the promise of a controversial set of measures consisting of shooting sulphate particles into the stratosphere, fertilising the oceans, or installing artificial trees that can suck carbon dioxide from the air. It is the symbolic moment when climate engineering materialised in the US political realm and assumed the form that continues to define the debate over climate engineering until today. In addition to mitigating and adapting to climate change, climate engineering became established as a potential third kind of policy approach to tackling climate change – 'a third possible risk-management strategy for climate change' – as one federal agency put it.[3]

APPROACHING A WATERSHED MOMENT

Before exploring this starting point further, let's take a step back. Fig. 2.1 may help to provide a bigger picture of where we stand at this moment in time. The graph situates the historical moment of Gordon's hearing within the wider context of US climate policy. It traces the rise in policy attention to climate

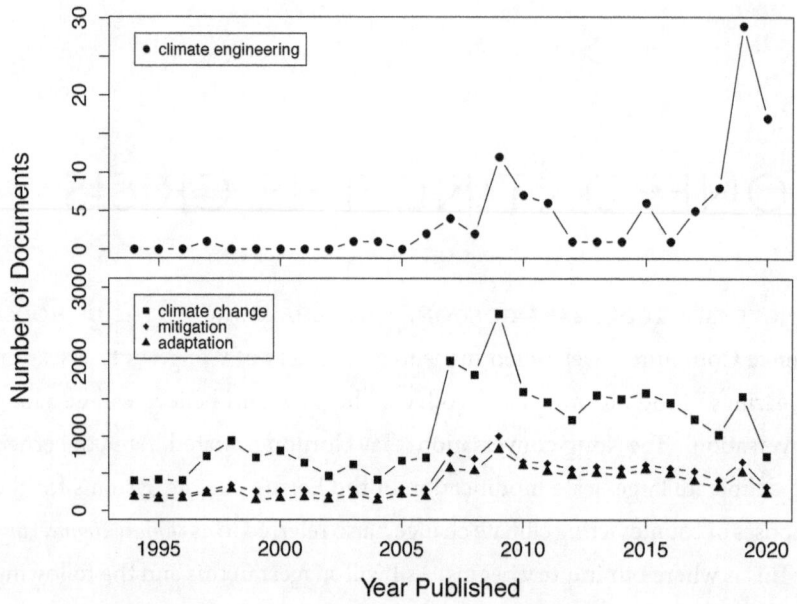

FIG. 2.1 Climate Engineering in US Climate Policy (FDsys)

Upper Panel: All policy documents addressing the topic of climate engineering (105 records in total) in the US Federal Digital System (FDsys) in the years from 1994 to 2020. Lower Panel: All policy documents addressing the topics of global climate change (squares, 29.383 records in total), the mitigation of global climate change (diamonds, 11.955 records in total), and the adaption to global climate change (triangles, 9.569 records in total) in the US Federal Digital System (FDsys) in the years from 1994 to 2020.

engineering for the period of 1994 to 2020 and compares this to discussions addressing other measures to counteract climate change.[4] The graph draws on a corpus of policy documents. It builds on records in the digital database of the US Government Publishing Office (FDsys). The squares in the lower panel display the distribution of all policy documents addressing climate change, with the other three plots charting trends in policy that attend to different modes of tackling this issue: mitigating climate change (diamonds, lower panel), adapting to its consequences (triangles, lower panel), and climate engineering (dots, upper panel).

The graph suggests that between the early 1990s and 2006, climate change slowly emerged as a relevant issue within US climate policy, accompanied by the continuous exploration of mitigation and adaptation as potential response measures. Climate engineering, by contrast, played virtually no role in these debates. The politicisation of climate change during these years thus did not stimulate political exploration of climate engineering as a potential response measure. In fact, climate engineering did not become an issue within US climate policy until the dawn of the new millennium, lagging notably behind as the chart begins to suggest. It is only in the period between 1997 and 2009 that these controversial measures slowly begin to appear on the political agenda. Still, however, political attention to climate engineering remains very cautious during these years. Between 1997 and 2006, the topic only pops up in a total of seven policy documents. Part of the reason for this dynamic is likely a general shift of attention away from the issue of climate change in the aftermath of the terrorist attacks of September 2001.[5]

Around 2006, however, the issue of climate change gained renewed traction, and this time, exploration of climate engineering followed suit. In 2009, US political attention to climate engineering reached its first peak. Document density jumped from two documents in 2008 to twelve in 2009. While this attention to climate engineering levelled off somewhat after 2009, ten years later, in 2019, it reached a second substantial peak. Document density jumped from eight relevant policy documents in 2018 to a total of 29 in 2019. And as the climate agenda of the incoming Biden administration suggests, it seems likely that climate engineering will only gain political traction in the years to come.[6] The issue, in other words, appears to be here to stay.

This growing political attention to climate engineering corresponded to growing funding levels for research and development, both from state and private sources, as well as intensifying publication activity in this field. While there is, to my knowledge, no comprehensive account of climate engineering funding allocated from 1994 through 2020, we gain a glimpse of the overall dynamic by combining several accounts and sources. Take solar radiation management, for example. The Harvard Solar Geoengineering Research Program suggests that government funding for this particular version of climate engineering has, so far, been miniscule, especially when compared to other climate related research. The group does, however, point to increases in funding over time. While there was a little over $1 million of funding in 2008, in 2018 this figure was a little over $8 million.[7] In 2016, the scientists noted a distinct spike in funding which can largely be attributed to the inception of two critical research programs, namely Harvard's Solar Geoengineering Research Program and the Carnegie Climate Geoengineering Governance Initiative.

More recent funding decisions suggest continuously growing political interest in the issue. In December 2019, for example, President Trump signed a spending package which earmarked a total of $4 million for solar radiation management related research. Congress directed these funds to a federal research agency – the National Oceanic and Atmospheric Administration (NOAA) – for studying, among other things, 'the impact of the introduction of material into the stratosphere from … proposals to inject material to affect climate, and the assessment of solar climate interventions'.[8] This decision attracted much media attention as the 'first time ever' that the US government 'allocated funding for a federal agency to conduct geoengineering research'.[9] In 2020, the House of Representatives proposed a 2021 budget for that same agency, NOAA, which, if enacted, would more than double these funds, adding another $5 million to the budget.[10] And in that same year, Silver Lining, a non-profit organisation based in Washington D.C. announced its Safe Climate Research Initiative (SCRI) 'to advance critical research in the historically underfunded field of solar climate intervention', promising yet more money for solar radiation management related research.[11] In a first step, the initiative awarded $3 million to a number of research programs not only in the United States, but also the United Kingdom and the 'Global South'.[12]

We can trace a similar dynamic with respect to carbon dioxide removal. While US government spending has been larger for this climate engineering approach, funds have increased dramatically since the early 2000s. An analysis by the Bipartisan Policy Center, a Washington-based think tank, estimated that in the decade between 2009 and 2019 government spending on direct air capture measures comprised $10.9 million in total.[13] For the 2020 fiscal year, Congress allocated twice this amount to the Department of Energy alone to research direct air capture and other negative emissions technologies.[14] And if enacted, the House appropriations bill for 2021 would triple this amount, directing a total of $40 million to the Department of Energy to study direct air capture measures and an additional $25 million to establish a 'Direct Air Capture Center'.[15] The US National Academies – one of the most authoritative national institutions dedicated to the advancement of science – recommended that funding levels for direct air capture should be ramped up even further, suggesting a budget of between $1,810 million and $2,400 million over a ten-year period.[16]

Analyses of publication output further complement this picture, helping us to contextualise this watershed moment in the career of climate engineering. In addition to growing political attention and increasing funding levels, bibliometric studies trace intensifying publication activity in the field of climate engineering, around 2009.[17] Oldham and others, for example, noted more than a tripling of publication output in 2008.[18] Belter and Seidel found that more than half of all climate engineering articles published between 1988 and 2011 have appeared since 2008. The authors suggest that a series of ocean fertilisation experiments between 1988 and 2008 seems to explain most publication activity between 2000 and 2008, while solar radiation management publications follow suit, displaying 'an exceptionally large increase' between 2006 and 2009.[19]

In self-descriptions of the field, this substantial boost in scientific attention to the topic is commonly pinned down to the publication of one distinct paper in 2006.[20] In this editorial essay, Paul Crutzen, Dutch atmospheric chemist and Nobel Laureate, asked if solar radiation management could provide 'A Contribution to Resolve a Policy Dilemma?'[21] The landmark essay, published in *Climatic Change* together with five commentaries on the text, managed to spark a heated debate along with 'harsh criticism'.[22] In hindsight, Crutzen has been praised for having lifted the 'taboo' on climate engineering and particularly solar radiation management research.[23] As a renowned scientist, particularly acclaimed for his contributions to ozone chemistry, he is seen as having effectively provided legitimacy to the controversial suggestion of engineering the climate.[24]

Taken together, this series of examples helps situate this moment in time in the career of climate engineering in a first approach. It begins to suggest just how substantially the dynamic of the climate engineering debate shifted its pace and quality around 2009, from rising political attention to the issue over increasing funding levels to intensifying publication activity in the field.

A BAD IDEA WHOSE TIME HAS COME

We can return now to Gordon's hearing and further delve into the 'long conversation' that he sought to initiate here. Gordon had placed climate engineering on the congressional agenda just a couple of weeks before the climate negotiations in Copenhagen. It was the 15th meeting of the Conference of the Parties to the United Nations Framework Convention on Climate Change (UNFCCC). The climate policy world had its hopes up and its eyes fixed on the Danish capital as delegates from around the world flocked to Copenhagen to try to negotiate a binding agreement on mitigating climate change. Fig. 2.1 suggests the distinct peak in US policy attention focused on global climate change during that year. While the initial goal of the meeting was a joint commitment to emissions reductions, delegates were left with nothing more than an informal agreement. The Conference ended in 'failure, rancor and disillusionment'.[25] The goal of effectively tackling climate change by mitigating greenhouse gases seemed to recede into the dim distance.

Yet, when Gordon opened his hearing on 'the deliberate large-scale modification of the Earth's climate systems for the purposes of counteracting climate change',[26] he wanted to make sure it had absolutely nothing to do with the pending negotiations in Copenhagen. Anticipating 'misleading headlines', he explicitly disconnected the hearing from the Copenhagen delegates' quest to commit the world to a global response to the problem of climate change. As we continue to follow Gordon's opening remarks, he seems to outright reject the very idea which his hearing's program was officially proposing, namely, to counteract climate change by means of the deliberate large-scale modification of the Earth's climate:

> But before we begin this discussion today, I want to make something very clear upfront. My decision to hold this hearing should not in any way be misconstrued as an endorsement of any geoengineering activity, and the timing has nothing to do with the pending negotiations in Copenhagen. I know we will run the risk of misleading headlines.[27]

A similar tone defines the testimonies of the invited expert witnesses, presenting, for example, lists of 'reasons why geoengineering might be a bad idea',[28] emphasising 'in the strongest possible terms' that they were 'not arguing that the US or anyone else should engage in [solar radiation management]',[29] that 'the United States Government should make it absolutely clear we are not planning for deployment of climate intervention technology',[30] and that 'climate engineering technologies do not now offer a viable response to global climate change'.[31]

Climate engineering, in other words, gained political traction during the early 2000s as 'a bad idea whose time has come', as science journalist Eli Kintisch put it.[32] The controversial measures successfully arrived on climate policy agendas as a kind of last resort option, a 'Plan B', something that needed to be considered, but would not really be an option and definitely not a solution against global warming.

While the 1980s and 1990s had established climate change as a complex, yet containable problem, the early 2000s witnessed a change of perspective and came with a surge of publications which emphasised the increasingly dangerous impacts of climate change on society. Metaphors of 'climate emergencies' and 'tipping points', from which there would be no return, had reached political and scientific attention.[33] This language marked a growing sense of urgency about how to tackle this problem. Climate change was experienced primarily through 'science-fuelled imagination'[34], but extreme weather events, images of melting icecaps and starving polar bears all added tangibility to the daunting crisis.[35] Indeed, as various scholars have pointed out, there was something of a 'crisification' of climate change in the years leading up to the 2009 climate negotiations in Copenhagen.[36] We will come back and unpack the role of particularly numerical modes of observing and problematising climate change to advance the notion of climate engineering in Chapter 6.

Against this backdrop, experts and policymakers alike had begun to argue for the need to look into climate engineering, not by drawing on positive images of techno-scientific innovation, but by invoking the impending climate crisis. This was when Paul Crutzen – the Nobel laureate chemist who we met earlier in this chapter, acclaimed for having 'lifted the taboo' on climate engineering – famously dismissed the desirable option of effectively counteracting global warming by the sole mitigation of greenhouse gases as 'a pious wish'.[37] Following the same line of reasoning, chairman Gordon motivated the Science Committee's inquiry into climate engineering by drawing on a 'stark'[38] and 'unfortunate reality'[39], namely that the '[...] onset of climate change impacts may outpace the world's political, technical, and economic capacities to prevent and adapt to them'.[40] The experts invited to the 2009 hearings similarly suggested that '[t]he problem is too serious to allow prejudice to take options off of the table'[41] and that it is therefore 'time to take the option [climate engineering] out of the closet'.[42] In these testimonies, the need for climate engineering research was deeply embedded in frightening scenarios such as this:

> What if we were to find out that parts of Greenland were sliding into the sea, and that sea-level might rise 10 feet by mid-century? [...] What if rainfall patterns shifted in a way that caused massive famines? What if our agricultural heartland turned into a perpetual dustbowl?[43]

Experts furthermore suggested that this dire situation directly concerns national security needs: 'direct intervention in the climate system might someday save lives and reduce suffering of American citizens'.[44] Phil Willis, then chairman of the UK Science Committee, also mirrored Crutzen's sentiment in his testimony, concluding that '[t]he decision not to consider any initiative other than Plan A – mitigation – could be considered negligent'.[45] He had therefore 'urged' the government of the UK 'to consider the full range of policy options for managing climate change', including 'various geoengineering options as potential Plan Bs, in the event that Plan A, mitigation and adaption, was not sufficient'.[46]

Gordon followed suit by formally recommending that 'comprehensive and multi-disciplinary climate engineering research at the federal level' should be considered 'as soon as possible' to be prepared 'for future climate events'. Going even further, he urged for the need of a policy consensus on what would constitute a 'climate emergency' that would legitimately warrant 'deployment of [solar radiation management] SRM technologies'.[47]

What these observations begin to suggest is that references to the daunting climate change catastrophe served to make the assessment and pursuit of this 'bad idea' consistent, rational, and legitimate.[48] And it is precisely this notion of *choicelessness* – the suggestion that, in the face of dangerous climate change, we simply cannot afford to ignore these controversial measures – that has, in the years following 2009, continued to successfully push the idea of climate engineering further into the political limelight. In 2015, for example, the US National Academies for Sciences pointed to the particular contemporary historical circumstances as a rationale for their landmark inquiry into climate engineering. The working group argued that the time had come to look into these measures because 'as a society, we have reached a point' where the 'risks from climate change' seem to 'outweigh' the risks of 'a suitably designed and governed research program' on climate engineering.[49]

Two years later, the Geoengineering Research Evaluation Act of 2017 was introduced into Congress, tasking the Academies with yet another report on climate engineering, and more specifically, with devising 'a research agenda to advance the understanding of albedo modification strategies'. The bill established the need for such a report by pointing, first, to 'the severe impacts' of global warming 'on human health, the global economy, and United States national security', to then argue for additional measures in tackling climate change: 'cutting carbon pollution is still the best way to mitigate climate change ... However, the United States and other nations may also need to consider climate intervention strategies'.[50]

Again, climate engineering appears in these observations as an undesirable, yet inevitable fate, as something that needs to be faced, whether we want to or not. That same year, the federal climate change research program (the US Global Change Research Program) suggested that in the face of the 'severely challenging task' of limiting the global mean temperature rise or adapting to the impacts of a warmer world, 'some scientists and policymakers' have shown 'increased interest ... in exploring additional measures' such as 'geoengineering or climate intervention (CI) actions'.[51] And in June 2020, the Select Committee on the Climate Crisis[52] argued that climate engineering was needed as a way of 'solving the climate crisis'.[53] The committee's Congressional Action Plan includes research and development on both solar radiation management and carbon dioxide removal measures. Carbon dioxide removal, particularly in the form of direct air capture measures, in fact provides a key component in this plan.[54] For the context of solar radiation management, the committee suggested that 'if global efforts to mitigate carbon emissions falter, and as the impacts of climate change continue to worsen, governments may consider alternative approaches to intervene in the atmospheric climate system'. Adding to the notion that we simply have no choice in dealing with these measures, the authors invoke a kind of techno-scientific arms race:

> Deploying ACI [atmospheric climate intervention] would be, at best, a modest complement. Nonetheless, the possibility of future deployment of ACI, including by foreign governments or non-state actors, necessitates consideration of the risks and governance of ACI.

Considering that climate engineering becomes a necessity, the committee recommended that 'Congress should ... establish a research program to investigate potential ACI approaches, their risks, and governance frameworks'.[55]

MAKING SENSE OF THE POLITICS OF CLIMATE ENGINEERING: CONFRONTING NARRATIVES OF CHOICELESSNESS

These actors' accounts thus suggest that climate engineering provides a rather paradoxical case of a techno-political project. Instead of resonating as a positive vision of socio-technical innovation, climate engineering gained political currency during the first decade of this millennium as an unappealing, even daunting measure of a last resort, a wholly disenchanted vision of 'science to the rescue'. Scholarship on the discursive framing of climate engineering has demonstrated how these measures became established during the early 2000s as a 'prudent' strategy of 'risk-reduction, management and control'.[56] The literature identifies 'a whole family of metaphors' that formulate climate engineering as an insurance strategy against the daunting climate catastrophe.[57] In these observations, climate change appears as a chronic disease,[58] a car or a plane crash,[59] against which climate engineering provides the potential remedy. In a conversation with the New Yorker in May of 2012, Hugh Hunt – an engineering professor at Cambridge University who was part of the first attempt to undertake an experiment on Stratospheric Particle Injection for Climate Engineering (SPICE) – perfectly captures this sentiment by suggesting:

> I know this is all unpleasant. [...] Nobody wants it, but nobody wants to put high doses of poisonous chemicals into their body, either. That is what chemotherapy is, though [...]. This is how I prefer to look at the possibility of engineering the climate. It isn't a cure for anything. But it could very well turn out to be the least bad option we are going to have.[60]

With regards to the debate in the media, Holly Buck finds that 'almost nobody' presented their accounts 'with attention to the positive power of humans to transform their societies or environments'.[61] After all, the success of this controversial idea seems intimately linked to the notion of choicelessness:

> Humans, even when they are cast as fixers, are rarely protagonists. Even the articles [on geoengineering] which featured ecological modernization were not exactly enthusiastic or positive: more often, they approached managing the earth as a chore, rather than a creative activity. The actors featured seem unable to act ... It is necessary to stabilize the climate to avert chaos – as Boykoff et al (2010: 60) explain, 'a guiding ethos of climate stabilization is the imagined future, safe, secure, stable climate, which can be engineered by our actions now'. Yet this stability is about averting the negative, not about establishing something positive.[62]

These perspectives, then, beg the question of how to make sense of this conflicted status of climate engineering. I want to suggest in this book that to engage in a meaningful way with this controversial debate over climate engineering and with the politics of techno-scientific innovation more broadly, we need to unpack and confront this notion of choicelessness. Referring to urgency in efforts to argue for the need to counteract climate change is of course important (the risks of climate change and extreme weather events are very real and constantly increasing). Urgency alone, however, fails to explain the necessity, not to mention the inevitability of climate engineering in contrast to other approaches of addressing the problem, such as radical emission cuts, political-economic reorganisation, or drastic changes in energy consumption and lifestyles.[63]

To provide a more satisfactory account of how we got here, we need to take this narrative of choicelessness seriously *as an actor's category*. This means that instead of arguing why climate engineering is or is not a Plan B,[64] we need to unpack how this techno-political project gained traction precisely as such. We need to place this understanding of climate engineering in its historical context, and more specifically, we need to understand its historical status in relating climate science to politics. The histories of climate engineering are necessarily manifold. This book seeks to complement accounts of climate engineering as an unprecedented last resort with an inquiry into the grown alliances that have driven political interest in climate science for decades. If we disentangle the science-politics interrelations that have shaped the recent arrival of climate engineering on the political agenda, we see that next to this temporal dynamic of crisis and fracture sits a story of continuity. This book aims to integrate these perspectives.

NOTES

1 Chairman Gordon in US House of Representatives, Committee on Science and Technology (2009: 11).
2 US House of Representatives, Committee on Science and Technology (2009: 3).
3 US Government Accountability Office (2010b: 2). See also US Government Accountability Office (2011).
4 This quantitative display of policy documents is limited to the years 1994–2020 as most of the relevant document collections within FDsys are only digitally available from 1994 onward. This applies, for example, to congressional hearings, the Federal Register, congressional bills, the Congressional Record, or congressional documents (see US Government Publishing Office (2018) for a detailed account of the availability of all FDsys collections). To explore the origins of climate engineering in US politics before that, we will turn to historical scholarship and scientific assessment reports (see Chapters 3 and 4).
5 See also, e.g., Brechin and Freeman (2004: 11f.); McCright and Dunlap (2011: 159); Turner and Isenberg (2018: 175).
6 See, e.g., The Biden Harris Campaign (2020).

7   Necheless and others (2018).
8   US House Committee on Appropriations (2019: 17–18); see also Schwarber (2020a); US House of Representatives, Select Committee on the Climate Crisis (2020: 526).
9   Pontecorvo (2020); see also Temple (2019).
10 Schwarber (2020b).
11 SilverLining (2020).
12 SilverLining (2020).
13 Hezir and others (2019: 6, 24).
14 Peterson (2020); see also US House of Representatives, Select Committee on the Climate Crisis (2020: 279).
15 Peterson (2020: 21); see also Bright (2020).
16 National Academies of Sciences (2019: 234–35, 246); see also Hezir and others (2019: 24).
17 See, e.g., Belter and Seidel (2013: 420); Oldham and others (2014); Linnér and Wibeck (2015: 257).
18 See Figure 3.1. in Stilgoe (2015: 186). See also Kintisch (2010: 12).
19 Belter and Seidel (2013: 420).
20 See, e.g., Keith (2013: 92); Morton (2016: 152f.) or Stilgoe (2015: 133ff.) for a critical perspective on this publication's importance.
21 Crutzen (2006).
22 Lawrence and Crutzen in Blackstock and Low (2019: 91).
23 Keith (2000: 92).
24 Hulme (2014: 4).
25 Hulme (2014: viii).
26 US House of Representatives, Committee on Science and Technology (2009: 3).
27 Chairman Gordon in US House of Representatives, Committee on Science and Technology (2009: 11).
28 Robock in US House of Representatives, Committee on Science and Technology (2009: 53f., 43); see also US House of Representatives, Committee on Science and Technology (2010b: 39).
29 Morgan in US House of Representatives, Committee on Science and Technology (2009: 276).
30 Long in US House of Representatives, Committee on Science and Technology (2009: 302).
31 Government Accountability Office (2011: i).
32 Kintisch (2010: 13).
33 Kintisch (2010: 39).
34 Hulme (2014: ix).

35 There is a host of literature, examining the implications and effects of different formats of climate change communication. For a review, see, e.g., Moser (2010). For a comparison between different communication formats (texts, charts, metaphors), see, e.g., van der Linden and others (2014).
36 See, e.g., Paglia (2018).
37 Crutzen (2006: 217).
38 Gordon in US House of Representatives, Committee on Science and Technology (2009: 11).
39 US House of Representatives, Committee on Science and Technology (2010b: III).
40 US House of Representatives, Committee on Science and Technology (2010b: III).
41 Caldeira in US House of Representatives, Committee on Science and Technology (2009:17).
42 Victor qtd. by Morgan in US House of Representatives, Committee on Science and Technology (2009: 293).
43 Caldeira in US House of Representatives, Committee on Science and Technology (2009: 21).
44 Caldeira in US House of Representatives, Committee on Science and Technology (2009: 21). See also, US House of Representatives, Committee on Science and Technology (2009: 11ff., 21, 43, 317).
45 Willis in US House of Representatives, Committee on Science and Technology (2009: 242).
46 Willis in US House of Representatives, Committee on Science and Technology (2009: 230). See also United Kingdom House of Commons.
47 US House of Representatives, Committee on Science and Technology (2010b: 37ff., 40).
48 See, e.g., Sheperd, Jackson in US House of Representatives, Committee on Science and Technology (2009: 28, 177).
49 US National Research Council (2015b: 10).
50 H.R.4586 (2017).
51 US Global Change Research Program (2017: 401).
52 The US House of Representatives, Select Committee on the Climate Crisis is a successor of the Select Committee on Energy Independence and Global Warming (2007–2011).
53 US House of Representatives, Select Committee on the Climate Crisis (2020).
54 See particularly US House of Representatives, Select Committee on the Climate Crisis (2020: 276–83).
55 US House of Representatives, Select Committee on the Climate Crisis (2020: 526).

56 Fragniere and Gardiner (2016).

57 See also Nerlich and Jaspal (2012); Huttunen and Hildén (2014); Luokkanen, Huttunen and Hildén (2014); Markusson, Ginn and others (2014); Kreuter (2015); Bellamy in: Blackstock and Low (2019: 51f.). For a critique of this emergency framing, see Horton (2015); Sillmann and others (2015); Fragniere and Gardiner (2016); Markusson, Gjefsen, and others (2017).

58 See, e.g., US House of Representatives, Committee on Science and Technology (2009: 153, 298). Or Hugh Hunt in Specter (2012).

59 Caldeira and Lane in US House of Representatives, Committee on Science and Technology (2009: 22, 37).

60 Specter (2012).

61 Buck (2013: 176).

62 Buck (2013: 176).

63 Scholarship on the history of environmental reflexivity and the Anthropocene has convincingly argued that popular narratives of the redemptive power of science in the climate crisis run the risk of leaving unattended human agency in making this crisis, in making sense of this crisis, and in addressing this crisis (see, e.g., Bonneuil and Fressoz (2016); Hulme (2014); Locher and Fressoz (2012).

64 Fragniere and Gardiner (2016).

2

THE EMERGING POLITICS OF CLIMATE ENGINEERING

IN 2009, CLIMATE ENGINEERING OFFICIALLY ARRIVED ON THE US CONGRESsional agenda. As we have seen in the previous sections, this is when climate engineering took shape as an issue in its own right, not only in US climate policy, but around the globe. But what exactly does this mean, we may ask, for a set of techno-scientific concepts that largely 'remain to be invented'?[1] How, in other words does a rather diffuse, even speculative, set of techno-scientific measures take concrete political shape? How does it unfold its political status and meaning?

This chapter sets the book's stage. It follows climate engineering through the lens of federal US policymaking. The chapter distinguishes two distinct *arenas* in which climate engineering began taking shape in the political sphere around 2009. On the one hand, climate engineering materialised as a matter of fact(s); it was assembled as a policy measure when policymakers and experts began establishing an 'official record' on the issue. On the other hand, climate engineering became structurally internalised by the political system; it took shape as a set of techno-scientific challenges that began guiding efforts to steer the development of relevant expert capacities within the federal infrastructure.

Both contexts can be understood as arenas in which science 'meets' politics. This means these arenas not only draw our attention to different dimensions of the emerging 'politics' of climate engineering, but they also introduce us to distinct sets of scientific experts and notions of expertise, as well as suggesting different timescales at play in assembling this techno-political project of climate engineering. In a nutshell, this chapter argues that understanding how climate engineering came to be requires asking how scientific struggles have come to 'match' political struggles.[2]

PRODUCING AN OFFICIAL RECORD, ESTABLISHING THE FACTS

To the public, the arrival of climate engineering in US politics became particularly visible around 2009, when a number of hearings, reports, and position statements on the issue began popping up. From the congressional inquiry and public hearings on the issue to the internationally discussed Royal Society report and the various volumes prepared by the US National Academies, it seemed like a pile of documents was being produced, each somehow seeking authority in defining climate engineering and determining its potential as a policy measure. Table 3.1 provides an overview of the most prominent of these documents that have informed the US political exploration of this issue, beginning in 2009.[3] In the following, I want to suggest that this pile of documents can be understood as a critical arena in which climate engineering began to take political shape.

Specifically, climate engineering took shape in these documents as a matter of fact(s). Hearings are held and reports are documented, as stated by Ann Keller, 'to communicate something publicly'. They are the result of efforts by policymakers and scientists to produce an 'official record' on the issue of climate engineering. This pile of documents thus opens a 'purposive arena' of communication.[4] These formats have provided policymakers, as well as selected experts, with a visible platform to purposefully establish climate engineering as a political issue in its own right. They have been essential in defining 'what climate engineering is' in terms of categorising, ordering, and assembling it as a governance object, engaging various expert groups in definitional struggles over what is at stake, and in determining its promise, risks, and potential as a viable policy measure to counteract climate change.[5] Via this 'official record' on climate engineering, we can thus trace how this option of governing climate change by engineering the Earth's climate system was envisioned, and how climate change was made legible to politics as an engineering challenge.

As we will see, science 'meets' politics in this arena in the form of a kind of 'staged advice'. This means that, somewhat paradoxically, scientific expertise becomes politically meaningful here precisely by suggesting a clear division from 'the politics' of climate engineering. One of the key dynamics we can observe in this arena of politicisation is the attempt to 'de-politicise' climate engineering.

NAME	AUTHOR	YEAR	FORMAT
Geoengineering the Climate: Science, Governance and Uncertainty	The Royal Society	2009	Scientific Assessment
Geoengineering Parts I, II, and III	US House Committee on Science and Technology, 111th Congress	2009–10	Congressional Inquiry
Climate Change: A Coordinated Strategy Could Focus Federal Geoengineering Research and Inform Governance Efforts	US Government Accountability Office	2010	Congressional Inquiry
Climate Change: Preliminary Observations on Geoengineering Science, Federal Efforts, and Governance Issues	US Government Accountability Office	2010	Congressional Inquiry
Memorandum: International Governance of Geoengineering	Congressional Research Service	2010	Congressional Inquiry
Engineering the Climate: Research Needs and Strategies for International Coordination	US House Committee on Science and Technology, 111th Congress	2010	Congressional Inquiry
Climate Engineering: Technical Status, Future Directions and Potential Responses	US Government Accountability Office	2011	Congressional Inquiry
Geoengineering: Governance and Technology Policy	Congressional Research Service	2013	Congressional Inquiry
Climate Intervention: Carbon Dioxide Removal and Reliable Sequestration	National Research Council (National Academies)	2015	Scientific Assessment
Climate Intervention: Reflecting Sunlight to Cool Earth	National Research Council (National Academies)	2015	Scientific Assessment
Geoengineering: Innovation, Research, and Technology	US House Committee on Science, Space, and Technology, One Hundred Fifteenth Congress	2017	Congressional Inquiry

TABLE 3.1 The 'Official Record' on Climate Engineering in US Climate Policy. List of most prominent hearings, reports, and documents, informing the official inquiry into climate engineering in US climate policy.

We already gained a glimpse into what this means in the introduction to this book. This form of staged advice effectively suggests that what we observe in these hearings and assessment reports is an independent foundation of facts, separated from the potential future political decisions on the issue that they might inform, all the while assuming the highly political role of defining possible, feasible, or desirable futures in this context.[6] Keller referred to this phenomenon as the 'politics of objective advice'.[7] We will come back to this notion of 'staged advice' in Chapter 6. In the following, we will first distinguish two important components of this official record on climate engineering – congressional fact gathering and scientific assessment reports. We will unpack how science comes to bear on politics within these distinct settings, before turning to the definitional struggles that are waged here.

Congressional fact gathering

At the heart of the effort to produce an official record on climate engineering stood a programmatic congressional inquiry into the issue beginning in the fall of 2009 (see Table 3.1). In its own words, the House Science Committee of the 111th Congress 'began a formal inquiry into the potential for geoengineering to be a tool of last resort in a much broader program of climate change mitigation and adaptation strategies'.[8] The visible centrepiece of this inquiry consisted of a group of congressional hearings, which we briefly looked at in the previous chapter. Under the leadership of Bart Gordon, the Science Committee held three hearings under the banner of 'Geoengineering: Parts I, II, and III' in 2009 and 2010. These three hearings were part of a cooperative endeavour between the US and the UK parliamentary bodies, seeking to coordinate their efforts in establishing an evidence base on the issue. According to Gordon, it was a meeting in April 2009 with MP Phil Willis, then Chair of the UK Science and Technology Committee, that gave the impetus for the US inquiry into climate engineering.[9] Beyond the hearings, this inquiry was driven by legislative assessments. These were assessment reports, compiled by the Science Committee itself, and by two congressional support bodies, the Congressional Research Service (CRS) and the Government Accountability Office (GAO) (see Table

3.1). Legislative branch agencies integrate scientific and political observations intra-organisationally – i.e. their assessments and policy analyses are formally geared towards the needs and concerns of the legislative branch. It was such a request from GAO (in this case for 'information on geoengineering'), that initiated the formal congressional inquiry into climate engineering in autumn 2009.[10]

In the United States, such congressional inquiries are essential in starting the legislative pursuit of any newly emerging issue. They are instrumental in forming a political agenda; they serve as a carefully assembled foundation of evidence that can be built upon at different points in the legislative process – for example, when crafting legislation, when attributing governmental funds, or when initiating programs that concern the newly emerging issue. Climate engineering thus takes political shape in this context of a congressional inquiry through the perspective of commissioned and invited expert voices. Like the scientific assessment bodies, which we will turn to below, these hearings and reports literally set the parameters of the debate to come. Congressional hearings provide a visible platform to establish a topic without yet a clear political stance having to be taken on the issue. This seems particularly relevant in the context of controversial issues, such as climate engineering.

Scientific assessment reports

Scientific assessment reports have provided a further essential component of efforts to produce an official record on climate engineering. In this context, policymakers task expert organisations beyond the federal bureaucracy with providing scientific assessments. Examples in the UK and US include, for example, the Royal Society and US National Academies, which provide scientific assessments in response to concrete inquiries or requests. These institutions pool scientific excellence – primarily drawn from universities – for distinct, problem-oriented analyses of specific topics. Publications such as the 2009 report by the Royal Society, the 2015 and 2019 volumes by the US National Academies, or the reports by the Intergovernmental Panel on Climate Change (IPCC), have been essential to the emerging politics of climate engineering, as they, similar to the legislative inquiry into climate engineering, literally define the issue at stake. And by doing so, these assessments crucially guide the political exploration of the issue. Aarti Gupta and Ina Möller have demonstrated that these assessments, in fact, 'constitute a source of de facto governance'. That is, they effectively exercise governance effects despite this being unacknowledged.[11] In addition to determining knowledge gaps and formulating research needs, these scientific assessments effectively structure 'de jure types of governance' by normalising and institutionalising the issue at hand.[12]

Throughout the congressional hearings on climate engineering, policymakers and expert witnesses have mobilised individual observations provided by the Royal Society, the US National Academies, and the IPCC as something akin to a baseline of accepted 'facts' on climate engineering. These reports became politically relevant not simply by uncovering hitherto unknown information, but by structuring the political inquiry; they essentially guided the debate, as policymakers and experts referred to these reports as providers of institutionally certified positions.[13]

Take, for example, the response to the testimony of John Shepherd, chairman of the Royal Society report, in the first programmatic congressional hearing on climate engineering in 2009.[14] Ever since Shepherd's appearance, the report's findings have been a key reference point in attempts to shape a universally accepted definition of climate engineering. The appraisal by such a prestigious scientific association has served as a critical source of political legitimacy in discussing these measures. The Royal Society report has assumed almost unrivalled prominence in structuring political exploration of climate engineering and linking scientific to political observations in this context. By doing so, the report effectively governed the further development of the overall field of climate engineering research, differentiating research communities, preparing research programs, and guiding funding streams.[15]

We can see similar impacts from the reports by the National Academies of Sciences, Engineering, and Medicine (NASEM) and the IPCC, which have continually guided US political explorations of climate engineering. The congressional inquiry into climate engineering has pointed to the almost ceremonial relevance of the IPCC in defining the official status of the climate change issue and suggesting legitimate response measures. The IPCC's decision to either include or exclude climate engineering concepts in its assessment reports have been closely monitored, with policymakers discussing the extent or scope to which this happened, the particular choice of working group or chapter that addresses these concepts, and the distinct choice of words used to frame its findings and position statements.[16] As we will see in later chapters, NASEM, too, has explored the notion of climate engineering in a range of reports published since 1992. Looking back, these reports document how substantially the shape of climate engineering policy has shifted over the years.[17]

Definitional struggles: Devising modes of climate intervention

The two settings of congressional fact-gathering and scientific assessments draw our attention to a variety of *definitional struggles* that have surrounded the production of this 'official record' on climate engineering, and which we will turn to in the following. These definitional struggles shed light on the contestation that is involved in the formal establishment of relevant 'facts', on the work of categorising, demarcating, ordering, and assembling climate engineering as a potential policy measure. These struggles concern what climate engineering *is* or rather what it *should be*; they determine what kind of solution these measures promise and in response to what kind of problem. As a result, these struggles calibrate how the relation of science and politics is envisioned.

Umbrella Terms

When climate engineering arrived on the congressional agenda in November of 2009, it was primarily discussed as 'geoengineering'.[18] With this choice of label, the Science Committee, as well as the Government Accountability Office and the Congressional Research Service (CRS) followed the British Royal Society, which had presented its report 'Geoengineering the Climate' with much fanfare and public attention just a couple of weeks earlier. Both the congressional inquiry and the Royal Society report defined geoengineering almost identically as the 'deliberate large-scale intervention in the Earth's climate system, in order to moderate global warming'.[19]

Since 2009, the picture has become more complex. In addition to the still popular term of 'geoengineering', we can trace a growing variety of concepts that have been devised and pitted against each other to determine particular agendas surrounding the idea to deliberately modify the climate. In 2010, Bart Gordon, who we met earlier in this chapter as the chairman of the Science Committee, argued that actually, 'climate engineering' would be the more meaningful term:

> [...] I feel that [geoengineering] does not accurately or fully convey the scale and intent of these proposals, and it may simply be confusing to many stakeholders unfamiliar with the subject. Therefore, for the purposes of clarity, facilitating public engagement, and acknowledging the seriousness of the task at hand, this report will use the term 'climate engineering' in lieu of 'geoengineering' going forward.[20]

With the choice of 'climate engineering', Gordon thus marked a political vision for these measures, rather than paying heed to scientific intricacies. He emphasised the decided purpose and intent of this inquiry – the serious and targeted effort to engineer the climate. The Government Accountability Office seems to have followed this terminological suggestion, switching from mainly using the term 'geoengineering' in its 2010 reports to 'climate engineering' in its 2011 technology assessment.[21]

Others took issue with this choice of words. When, a couple of years later, the National Academies (NASEM) provided their 2015 study on the issue, they argued against the label of 'climate engineering' precisely for its misguided idea of control. The label would imply 'a greater level of precision and control than might be possible'. NASEM, however, also rejected the notion of 'geoengineering', as the concept would suggest 'a broad range of activities beyond climate (e.g., geological engineering)'. Instead, the experts suggested yet another label, namely 'climate intervention':

> The committee concluded that 'climate intervention', with its connotation of 'an action intended to improve a situation', most accurately describes the strategies covered in these two volumes.[22]

ENGINEERING THE CLIMATE

Ever since, this notion of 'climate intervention' or 'atmospheric climate intervention' has gained popularity, not only in scientific assessments.[23] Recent suggestions to speak of 'climate repair', 'climate restoration', or 'climate remediation' emphasise this intention of 'improvement' as suggested by the 2015 NASEM report even more explicitly, alluding to the ethical responsibility of humans to restore what has been harmed.[24] We will take a closer look at these labels in Chapter 6.

Distinguishing Modes of Intervention

Aside from arguing for a terminological adjustment, the 2015 NASEM report furthermore suggested a differentiation of the technical debates over climate engineering. The working group decided against one comprehensive study. Instead, the report came in two volumes, following two different kinds of climate intervention – either reflecting sunlight back to space or removing carbon dioxide from the atmosphere (see Fig. 3.1).[25] The Academies therein further reinforced the differentiation of climate engineering research as already suggested in the 2009 Royal Society report.

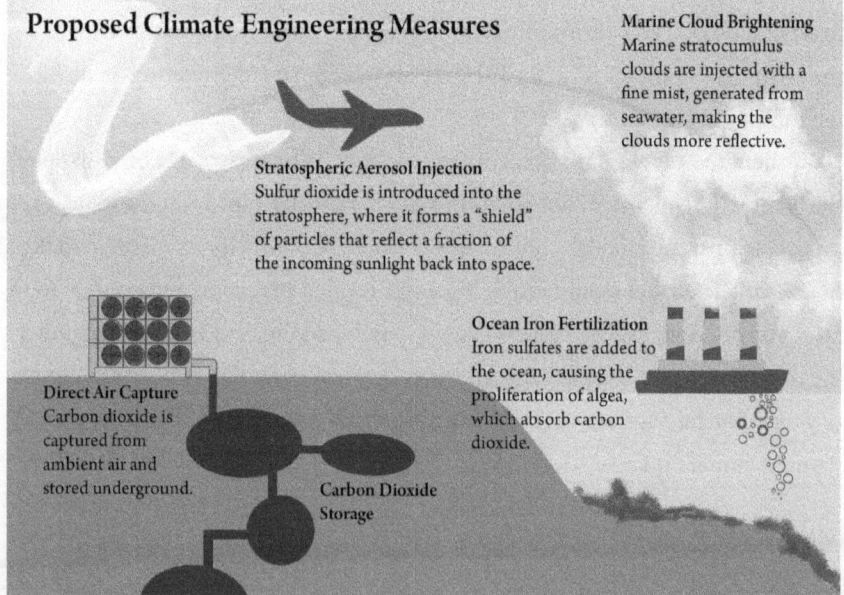

FIG. 3.1 Climate Engineering Proposals

Efforts to govern climate change by removing CO_2 from the atmosphere are generally referred to as carbon dioxide removal (CDR).[26] The boundary between what is considered as CDR – and therefore as 'climate engineering' – and what is considered as climate change 'mitigation', i.e. reducing anthropogenic impacts on climate, is somewhat fuzzy. There is no universally agreed upon demarcation between the two approaches and CDR measures are increasingly considered a key component in an effective mitigation strategy. However, a common heuristic in this context is to follow the source from which CO_2 is captured. The Royal Society suggests that any measure which captures CO_2 that has already been emitted into the atmosphere should be referred to as climate engineering.[27] Two approaches have gained particular attention in this context, both of which will play a prominent role in this book: so-called direct air capture (DAC) and ocean fertilisation measures.

Direct air capture describes various proposals to chemically 'scrub CO_2 directly from ambient air'.[28] The idea is to either use this extracted CO_2 to produce a concentrated stream of gas that can then be utilised in industrial processes or in the chemical production of carbon-based products, or to remove it from the air and permanently store it in a reservoir. For the long-term storage of the captured CO_2, various options have been discussed, such as utilising enhanced CO_2 mineralisation processes, like enhanced weathering, or storage in geologic formations.[29]

In addition to such land-based approaches, the discussion over CDR also includes ocean-based techniques. So-called ocean fertilisation measures seek to enhance the 'marine biological pump', that is, the natural carbon sink of the ocean.[30] According to NASEM, the basic idea is to stimulate growth in phytoplankton by adding limiting nutrients to surface waters and therefore increase the flow of organic carbon into the deep ocean.[31] The technique, which is most prominently discussed in this context, is the fertilisation with iron.

These CDR approaches are categorically distinguished from so-called solar radiation management (SRM) approaches. SRM generally refers to the idea of governing climate change by reflecting some of the incoming sunlight back to space.[32] The measures under discussion aim to enhance the Earth's reflectivity, which is also referred to as the Earth's 'albedo'. We find a multitude of labels used to describe such efforts in the current debate over climate engineering, some emphasising technical intricacies, some the political ambitions of these approaches. These include solar radiation management, solar geoengineering, sunlight reflection methods, albedo modification and many more.[33]

Two approaches have received particular attention in this context – marine cloud brightening and stratospheric aerosol injection. The former involves brightening marine stratus clouds by spraying seawater into the lower atmosphere. The basic idea is that seeding these marine clouds with tiny droplets of seawater would increase the cloud droplet concentration and enhance their longevity, meaning it would make them brighter and longer lasting, thus reflecting a higher fraction of incoming sunlight back to space.[34] The idea of Stratospheric Aerosol Injection, in contrast, is to replicate and technologically 'enhance' volcanic eruptions. The concept would thus entail injecting millions of tons of reflective particles into the stratosphere, where they are expected to remain for a longer time, forming a sun shield for the Earth.[35]

These visions of governing climate change by reflecting sunlight back to space or by removing carbon dioxide from the atmosphere have confronted science and politics with a set of concrete technical challenges. David Keith, one of the invited congressional experts, famously suggested that solar radiation management 'is cheap, fast, and imperfect',[36] while carbon dioxide removal has generally been assessed as comparatively safe, yet uneconomical and difficult to scale. In other words, what makes these two modes of climate intervention into two fundamentally different approaches is not merely their underlying 'science', but the effort to translate abstract scientific ideas into concrete policy measures. Both approaches come with distinct sets of technical challenges which arise precisely from efforts to match scientific observations on intervening in climatic change to the political challenge of governing climate change – and that means the challenge to provide feasible, effective, and safe means to tackle climate change.

In technical debates over climate engineering, experts' judgements over the status of the respective measures revolve around two different dimensions, as we will see in the following. On the one hand, their judgements concern what we might refer to as the political economy of climate engineering. That is, the political feasibility of the discussed measures, their status as viable policy tools. This is usually determined along the lines of cost, scalability or effectiveness, and risks or potential side effects of the devised measures. To put it differently, what hurdles need to be overcome to turn these approaches into economic, effective, and safe policy tools? On the other hand, the experts' judgements concern technical challenges or what we might refer to as the politics of evidence. That is, the grounds on which to judge the political feasibility and technological readiness or general status quo of the discussed approaches. We briefly touched on these issues in the introduction already, when approaching the question of what climate engineering *is*. The controversial debates surrounding climate engineering research governance nicely illustrate the epistemological battleground that lingers behind this line of research: are we looking at 'the first' experiment that 'tests a way to cool Earth' or at a harmless scientific exercise that merely squirts out some innocuous fluids?[37] How would we know? Where should we draw the line? How will it matter? Discussions in this dimension focus on questions of what can be considered as (politically) relevant, viable, robust forms of evidence; they concern the epistemological status of experimental or theoretical findings, of 'indoor' vs. 'outdoor' observations, of insights from modelling studies and computer simulations vs. insights from 'natural' analogies.

Governing Climate Change by Sucking Carbon from the Air

The 'official record' on efforts to govern climate change by removing carbon dioxide from the atmosphere is built on rather confident statements regarding the relevant evidence base. Congress, for example, found that the basic engineering principles of the discussed measures were comparatively well understood,[38] and the National Academies, too, judged that carbon dioxide removal would most likely 'not introduce novel global risks'.[39] Rather than unanticipated risks and side-effects, the experts deemed cost as a critical hurdle in realising CDR at scale ('at scale' means to an extent which would actually provide a meaningful policy tool for counteracting climate change).[40] According to the expert assessments, the political feasibility of CDR thus hinges primarily on questions of economic feasibility and commercialisation. In its 2020 Congressional Action Plan, the Select Committee on the Climate Crisis, for example, argued that carbon removal 'at a scale of 10 gigatons of carbon dioxide each year by midcentury' would be needed, while the largest operating plant in North America, a pilot plant by the Canadian company, Carbon Engineering, is able to remove a ton of CO_2 per day.[41] Drawing on the National Academies' assessments, the committee therefore emphasised that Congress would need to 'prioritise' direct air capture research and development in federal agencies.[42] Since 2019, Carbon Engineering has been working to engineer the world's biggest direct air capture plant in the world – a plant that, according to the company, is expected to remove one million tons of CO_2 per year (see figure 3.2.).

FIG. 3.2 Virtual Rendering of What Carbon Engineering's Large-scale Direct Air Capture Plants Will Look Like (Credit: Carbon Engineering Ltd.)

So, how does scientific expertise come to bear on the politics of climate engineering here? On the surface, it seems as if expert judgements on the feasibility of climate engineering exemplify a linear relationship between science and politics: in the case of carbon dioxide removal, the invited experts were able to build their assessments on quite a substantial body of evidence – a body of evidence that contains more than three decades worth of field trials, experiments, and demonstration facilities. Yet, this body of scientific research, its scope and quality, is hardly independent from political judgements on carbon dioxide removal. Instead, it rests – at least partially – on the political support for these measures, driven by interest from the fossil fuel industry.[43] The research itself, in other words, is importantly shaped by politics, both through available funding and existing regulation.

As a result of this comparatively large body of research and field studies, carbon dioxide removal, and specifically ocean fertilisation measures, remain the only climate engineering approaches thus far for which a regulatory framework exists, one that addresses, 'in principle', both research and implementation.[44] With ocean fertilisation field studies mounting since the early 1990s, political pressure from environmental NGOs and policy-oriented studies on the subject, for example, led to a 2008 resolution by the London Convention and Protocol, banning ocean fertilisation efforts for commercial purposes. Scientifically motivated field studies, however, are not prohibited, only subject to strict assessment.[45] The research landscape is thus shaped both by political support, as well as concerns and restrictions. Regulation seems to evolve in this case not prior to, but alongside with, research.

Although it may seem (and is presented) as though this 'official record' on climate engineering provides a neutral baseline of facts and figures for policymakers to base their decisions on, we can see here that the picture is much more complicated than that. If we ask for the genesis of the presented facts and their essential breeding ground, we readily see that politics is already involved in the very production of this 'official record'. The definitional struggles and experts' disputes that surround the establishment of this record nicely hint at the interdependence between the scientific research on climate engineering on the one hand, and the political interest in them, on the other. In other words, the very foundation of 'facts' that the political judgement on these technological approaches rest on, is itself a result of political judgements; the two are reciprocally coupled.

Governing Climate Change by Reflecting Sunlight Back to Space

Experts' judgements on the status of solar radiation management have looked quite different to those on carbon dioxide removal. Instead of cost and scale, it was in this case the associated risks and side effects that stood at the centre of the debate. Again, we find strikingly confident assumptions that solar radiation management will be 'cheap' and 'fast' compared to the reduction of emissions.[46] Such judgements established solar radiation management as a politically attractive tool that might be implemented swiftly and with large impacts,[47] all the while raising controversial debates over questions of governance. These questions included, for example, who would get to decide when to deploy and when to cease a potential SRM program, or how the legal liability of potential damages would be decided on. The official record on solar radiation management has therefore much more explicitly concerned the legitimacy and desirability of these approaches as a viable response to tackling climate change. Experts' assessments have warned quite prominently, for example, that solar radiation management would not address the causes of climate change, but only its symptoms.[48] Building on this observation, the so-called 'moral hazard' argument has played a crucial role in the debate over SRM. This argument holds that policymakers need to take into account the effect that the mere consideration of these measures has on the ultimate goal of climate change mitigation. The concern is that even the very idea of a potential 'sun shield' might make people feel 'insulated' from the risk of global warming, thus making them 'more likely to engage in risky or detrimental behavior'.[49]

In the case of solar radiation management, such normative and ethical questions regarding the basic outlook of these technological concepts and their potential risks and side effects were linked to epistemological concerns over how to decide on these issues. The central problem that the experts presented was how to examine the global effects and risks of a large-scale introduction of aerosol particles in the stratosphere.[50] They primarily disputed if and how technological effectiveness could be tested without actually deploying these measures, raising the question of how to generate a reliable and robust evidence base on the promise of SRM as a policy device.[51] Many of the invited experts emphasised the need for field studies to gather the kind of scientific evidence that would be necessary to avoid 'expos[ing] the world to serious risk' in the case of a sudden future 'emergency'.[52] They bound the possibility of a meaningful political decision on these measures at some point in the future to the need of field studies today.[53] The experts warned that essential engineering details, from ideal particle size to delivery mechanism – as well as potential side effects and risks – have remained understudied and undertested:

> How do we deliver the source to the region of release? [...] Once we have a detailed idea of precisely what source we want, can we produce that source? [...] After injecting the source in the stratosphere do particles form as models suggest? How do we track the plume? What instruments are required to measure the particle properties, the plume extent, and the reduction in sunlight below the plume. Do the particles coagulate and grow as our models suggest? Do the particles mix and evolve the way our models tell us they will (from source to the first scale, and from the first scale to the globe scale?).[54]

Yet, the question of how such a meaningful field study on solar radiation management would look was heavily disputed. How to *experiment* with altering incoming sunlight without actually altering incoming sunlight? While some experts argued that there is, in fact, a viable distinction between 'small-scale field studies' and 'full-scale deployment', others questioned this very distinction.[55] 'We are caught between a rock and a hard place', one expert witness explained:

> Too small a field test, and it won't reveal all the subtleties of the way the aerosols will behave at full deployment. A bigger field test to identify the way the aerosols will behave when they are concentrated will have an effect on the planet's climate [...]. I have not seen a suggestion on how to avoid this issue.[56]

To illustrate the dilemma, another expert witness compared the challenge of generating robust evidence on solar radiation management approaches to the historical process of understanding global warming:

[...] a real-world geoengineering experiment would have to be conducted for a long time, 10 or 20 years or longer, so as to gather enough data to calculate the statistics. It is only after 60 years of global warming since about 1950 and decades of the IPCC process that we have a clear understanding the greenhouse gases are responsible.[57]

As a result of this basic dilemma, the official record on solar radiation management crisscrossed a gaping divide between concrete assumptions regarding the problematic or promising ('cheap', 'effective') future reality of this 'technology' and the most basic epistemological questions concerning its design and effects. The scientific assessments and congressional inquiry connected sweeping and fundamental legal-normative questions to basic scientific challenges. They bound the task of anticipating the potential geopolitical consequences of a rogue state wielding an imaginary sun shield to the epistemological status of climate models; these assessments linked the normative and moral consequences of a speculative global thermostat to observations of a tethered balloon.[58]

Coming back to the 2015 NASEM climate intervention volumes, these assessments thus further cemented a differentiation of scientific debates along concerns over the policy implications and societal risks of the respective intervention approaches – namely CDR and SRM. NASEM argued that

> the committee's very different posture concerning the currently known risks of carbon dioxide removal as compared with albedo modification was a primary motivation for separating these climate engineering topics into two separate volumes.[59]

While the two-volume structure of the report reflects technical criteria concerning potential intervention approaches, the Academies' explanation for 'climate intervention' as a meaningful label reflects an orientation towards the broader societal vision of this research. It is this broader vision, then, which keeps a set of otherwise disparate lines of research under one roof.

This differentiation of climate engineering along CDR and SRM became further institutionalised when, in 2019 and 2020, NASEM began to prepare entirely separate reports on each set of the suggested intervention approaches, drawing on different groups of experts.[60] These assessments document an increasingly specialised policy debate that has begun moving away from umbrella terms of climate engineering or geoengineering, and instead fosters notions of 'negative emissions', 'atmospheric interventions' or 'sunlight reflection' measures. As we follow the technical details of these approaches, we encounter similar varieties of labels, each carrying layers of definitional struggles, rebranded purposes, and signs of the times.

As I suggested in the introduction to this book, these definitional struggles over climate engineering become meaningful as they demonstrate how the very concept of climate engineering, in all its semantic variations and evolutions, essentially bundles very different research contexts in their promise to fundamentally alter the politics of climate change. These definitional struggles precisely illustrate that the question of what climate engineering is and can be, is hardly just an academic one. The various labels and categorisations of climate engineering become meaningful and subject to heated debate as they bind various lines of scientific inquiry to the societal challenge of tackling climate change. The conceptual and semantic justifications for one against the other concept, calibrate this relation of scientific inquiry and political intervention, each emphasising different sides or aspects of this charged relation.

INTERNALISING CLIMATE ENGINEERING IN THE FEDERAL INFRASTRUCTURE

The official record on climate engineering – the established 'facts' on this matter – opened another arena in which climate engineering began to materialise in the political realm. The visions and assessments of what climate engineering is, can, or should be, presented politics with a concrete set of techno-scientific challenges, as we have seen in the previous section. These techno-scientific challenges did not merely remain the subject of abstract expert talk but had structural consequences too: politics began to structurally internalise climate engineering in the form of these techno-scientific challenges. It translated these expert assessments into federal programs, funding decisions, rules and legislation.

This arena of the emerging politics of climate engineering thus directly builds on the previous one. Yet, it draws our attention to a slightly different notion of 'politics'. Instead of bringing into focus the epistemic authority of selected experts and policymakers to assemble climate engineering as a governance object, this second arena sheds light on the structural consequences of these definitional struggles. It traces the political institutionalisation of this newly defined issue within the federal bureaucracy.

Science 'meets' politics in this arena not in the form of staged advice, but in the form of relevant expert capacities within the federal infrastructure. Policymakers in this case assessed which kinds of climate engineering-relevant expertise was already at their disposal within the federal bureaucracy. Building on this inventory, they took matters into their own hands, seeking to steer the proper establishment of such expert capacities and the respective advancement of research and development.

Taking inventory, charting new territory

One aspect of this political internalisation of climate engineering appeared as a kind of climate policy introspective. As part of its programmatic inquiry into climate engineering in 2009, Congress began to screen and inventory the federal landscape for relevant climate engineering expertise.[61] Around that same time, a number of initiatives and task forces appeared on the political scene, devising 'strategic plans' and charting roadmaps for policymakers and the government to advance and guide a future approach to the issue.[62] According to the Science Committee itself, the goal of this political inventory was to look forward. Such an inventory would allow policymakers to effectively 'guide future government and academic structures for research and development activities in this field'.[63] The arrival of climate engineering on the political agenda entailed the introduction of a new category in this context, a category, along which existing expert capacities and legal frameworks could be inventoried, assessed and developed. As a new category, climate engineering began re-aligning climate science and politics in the form of political resources and funds, expert capacities, and legal frameworks. To politics, climate engineering emerged here as both, already existing and entirely new.

We get an idea of how climate engineering took political shape in this case – how this 'inventory' bound past to future, the existing to the new – by turning to the assessment provided by the Government Accountability Office (GAO). As we have seen in the previous section, the congressional request for GAO's assessment formally introduced climate engineering to the US climate policy agenda in the fall of 2009.[64] This assessment concerned both existing research and development capacities, as well as the applicability of federal laws and legal frameworks to climate engineering.

To begin with, the Science Committee tasked GAO with assessing 'the extent to which the federal government is sponsoring or participating in geoengineering research or deployment'.[65] We can see how difficult and necessarily messy this endeavour must have been. How to decide what qualifies as a climate engineering and what does not? GAO decided to distinguish between three forms of climate engineering-relevant 'activities': (1) activities that were technically designated to 'conventional carbon mitigation efforts', but were, in fact, 'directly applicable to a proposed geoengineering approach'; (2) activities that concerned 'basic scientific understanding of earth systems, processes, or technologies' but might be 'applied generally to geoengineering'; (3) activities which are explicitly designated as climate engineering-relevant and do *not overlap* with a conventional carbon mitigation strategy'.[66]

The (albeit small) US budget for climate engineering in the 2009 and 2010 fiscal years varies substantially depending on whether all the above categories are considered as climate engineering relevant activities or only the explicitly designated ones:

> GAO's analysis found that 43 activities, totaling about $99 million, focused either on mitigation strategies or basic science. Most of the research focused on mitigation efforts, such as geological sequestration of CO2, which were identified as relevant to CDR approaches but not designed to address them directly. GAO found that nine activities, totaling about $1.9 million, directly investigated SRM or less conventional CDR approaches.[67]

As a result, the agency argued that, yes, 'federal agencies are sponsoring research relevant to geoengineering, but there is no coordinated federal strategy, making it difficult to determine the extent of relevant research'.[68] GAO presented its final results to Congress in 2010, pressing its main message already in the study's title: 'A Coordinated Strategy Could Focus Federal Geoengineering Research and Inform Governance Efforts'.

Congress additionally asked the Government Accountability Office for 'the extent to which federal laws and international agreements apply to geoengineering'.[69] This not only concerned the provision of relevant scientific expertise; it also meant evaluating the legal and regulatory frameworks within which this new approach to tackling climate change would operate. Formulating the challenge of governing climate change as a challenge of deliberately modifying climate change meant a substantial shift of perspective in this context. Climate engineering essentially turns the politics of climate change upside down.

In addition to commissioning the report by GAO, the US Science Committee dedicated the final of its three programmatic hearings on climate engineering especially to legal and governance questions, and it tasked the Congressional Research Service (CRS) with providing further advice on these issues.[70] In their assessments, both GAO and CRS examined the applicability of US law to this new category within US climate policy. The agencies took stock of existing entities and frameworks within the federal bureaucracy 'that might apply if climate engineering were tested or deployed at a large scale'.[71] Due to the global impacts of any envisioned climate engineering scheme, both agencies furthermore pointed to the relevance of international legal frameworks.[72] Against this backdrop, existing regulatory frameworks, such as ENMOD, the London Protocol, the Law of the Sea, the Antarctic Treaty, or the UNFCC were assessed regarding their applicability and their potential for respective adjustments.[73]

GAO remained rather vague in their assessment, simply noting that 'the extent to which existing federal laws and international agreements apply to geoengineering is unclear, and experts and officials identified governance challenges'.[74] The judgement of CRS illustrates the difficulties of translating an umbrella term, such as climate engineering, into concrete governance measures. The agency concluded that a flexible governance system would be needed as 'different technologies, different stages of the research and deployment cycle, and different environments for research and deployment activities may require different methods for oversight'.[75]

Tasking agencies with conducting original research

Aside from the US political inventory of relevant climate engineering capacities, the political system began to take matters into its own hands. In the following, we will see how politics internalised climate engineering into the federal infrastructure by approaching relevant agencies and departments directly and seeking to deliberately steer the development of climate engineering expertise. The prospect of advancing climate engineering as a potential policy tool turned basic scientific challenges into direct policy concerns; it made the political system attempt to jump scientific hurdles.

To begin with, the political system began internalising climate engineering by tasking federal agencies with conducting original research. In particular, the National Oceanic and Atmospheric Administration (NOAA), an agency within the Department of Commerce, emerged as a critical node of policy-relevant climate engineering expertise in this context. The agency was the target of both efforts to establish relevant carbon dioxide removal, as well as solar radiation management expertise.

We get a sense of what this means by examining the presidential proposals and congressional negotiations of the agency's budget between 2009 and 2012. In 2009, the House Committee on Appropriations argued that 'ocean fertilisation', one particular approach to carbon dioxide removal, '[…] has the potential to be used for climate change mitigation in the future, but that further research is needed' and that 'the Committee [therefore] encourages NOAA to support research into carbon sinks through ocean fertilization'.[76] One year later, the topic was again on the agenda of the agency's budget hearings. When the Science Committee assessed the Obama administration's budget proposals for NOAA, it wanted to know what research would be needed to better understand climate engineering, and specifically, what kinds of 'research capabilities, both internal to the agency and through external partnerships' NOAA could provide to contribute to such a better understanding.[77] Jane Lubchenco, then NOAA Administrator, pointed out that a successful climate engineering approach 'would require full scientific understanding of the underlying physical and chemical processes'. She emphasised the need for extensive research, not only natural scientific, but also economic, and suggested that 'enhanced communication and expanded efforts' between 'NOAA, other parts of the federal government, university and industry partners, and the international community' would become necessary.[78]

In its 2012 budget proposal for the agency, the House of Representatives officially tasked NOAA with assessing the mitigation potential of ocean fertilisation measures.[79] Specifically, the House Appropriations Committee suggested that the agency should 'address key scientific questions regarding the potential impacts of iron fertilization on the oceans' and should coordinate in this effort 'with other Federal agencies, academia, and the private sector, as appropriate'.[80] In the report that NOAA published in response to this inquiry, the agency suggested the immediate scientific merit of exploring this policy option. It reported that ocean fertilisation research

> has made an extremely valuable contribution to the scientific understanding of the ocean carbon cycle and its role in the global carbon cycle on time scales ranging from glacial episodes thousands of years in the Earth's past to today's changing climate.[81]

These records thus suggest how climate engineering effectively 'matched' political with scientific challenges, to use Zeke Baker's words here.[82] It directly and seamlessly bound political efforts at tackling a societal issue to scientific struggles of understanding the ocean carbon cycle. The political vision of governing climate change by removing CO_2 from the atmosphere made complex scientific challenges – challenges no less than gaining 'full scientific understanding' of the physics and chemistry of climate change – into an immediate political concern.

Almost ten years later, we can trace a similar dynamic with regards to solar radiation management approaches. In December of 2019, Congressman Jerry McNerney (D-California) introduced the Atmospheric Climate Intervention Research Act to the House of Representatives. This proposed legislation sought to amend the America COMPETES Act[83], a bill which was originally introduced by Bart Gordon and signed into law by President George W. Bush in 2007 to 'improve the competitiveness of the United States' by means of 'invest[ing] in innovation through research and development'.[84] If enacted, the proposed Atmospheric Climate Intervention Research Act would essentially formulate research into solar radiation management as one critical component of such a research program geared at national competitiveness and innovative capacities. Specifically, it argues that the prospect of 'inject[ing] material to temporarily reduce global radiative forcing of climate' introduces 'significant risks', which need to be properly monitored. It also adds that NOAA is responsible for that task.[85] The bill suggests that this task would require

> significant improvements to observations of the abundances and chemistry of the stratospheric gases and particles and the reflectivity of the stratosphere to establish the baseline state of the stratosphere and its trend over time and to develop enhancements to stratospheric models used for predicting climate impacts of material introduced into the stratosphere by natural or other means.[86]

Building on this assessment, the bill tasks NOAA with improving these observational and measurement capabilities so that it could provide an understanding of the 'proposed atmospheric interventions in Earth's climate' and particularly 'the effects of proposed interventions in the stratosphere and in cloud-aerosol processes'.[87] In effect, the bill thus formulates climate engineering as another critical chapter in the national strategic cultivation of climatological research. Solar radiation management emerges here not as a shocking change of perspective, but rather as a critical building block in the logical continuation of US political efforts to advance climate change expertise within the state.

Awarding cash prizes to push commercialisation of DAC

Aside from these efforts to steer the development of basic expertise within the federal bureaucracy, the political system also began to internalise climate engineering in the form of more concrete technical challenges. One example, which has gained a lot of political traction since 2009, is the commercialisation of direct air capture (DAC) technology. Expert assessments on the political feasibility of DAC have hinged primarily on questions of economic feasibility and commercialisation, as we saw in the previous section. The political system began translating and internalising this issue of economic feasibility and commercialisation of DAC in a number of ways, whether by authorising funds to advance original research within the federal bureaucracy, or by incentivising external research via Cash Prizes or grants, or by investing in demonstration facilities.[88] The political system in a sense sought to 'jump start a DAC industry', as the Climate Crisis Select Committee put it.[89]

I want to elaborate on just one example in particular: between 2009 and 2019, John Barasso, a Republican Senator from Wyoming, introduced a number of bills to Congress that sought to push commercialisation of direct air capture (DAC) technology by awarding cash prizes.[90] While the Carbon Dioxide Capture Technology Act of 2009 and the Carbon Dioxide Capture Technology Prize Act of 2011 were not enacted, the Utilizing Significant Emissions with Innovative Technologies Act of 2019 – the USE IT Act, for short – passed the Senate in 2019. Building on the technical possibility 'to separate carbon dioxide from … the atmosphere', the first two bills sought to 'provide incentives to the development and implementation' of technologies which would achieve this separation 'in an economical manner'.[91] The 2019 bill similarly emphasised that 'high cost' remains the 'main prohibitive aspect' when it comes to DAC technology.[92] But, how might the political system seek to steer techno-scientific innovation? In this case, it resorted to promising cash prizes. If enacted, the 2019 bill would authorise the administrator of the Environmental Protection Agency to administer a competitive prize program that awards 'up to $35 million in funding' to DAC research projects.[93]

As in the previous examples, the outlook of climate engineering essentially turns a scientific puzzle into a concern of direct national strategic relevance here. Alluding to the research project that eventually led to the building of the Atomic Bomb, the commercialisation of DAC technology was presented as a national priority in the 'Land of the Manhattan Project'.[94] We will come back to how important this framing of climate change as a national-strategic innovation challenge has been to the political advancement of climate engineering measures in the US since the early 2000s in Chapter 5. This language indicates how the outlook of climate engineering corresponds to a somewhat odd or at least striking problematisation of climate change. Climate change appears here not merely as a curious scientific puzzle, but as an innovation challenge that needs to be tackled by a concerted national effort. These prizes frame, define, and institutionalise climate change as a challenge that concerns techno-scientific innovation capacities as a matter of national pride and security. The labels assigned to the cash prize initiatives (on which the 2019 bill builds) invoke the transformative power of techno-scientific innovation for national strategic goals. These include 'Grand Challenges Prizes', 'Freedom Prizes', or 'Bright Tomorrow Lighting Prizes'.[95] Advancing research and development in this context emerges as an essential component of a national energy policy strategy,[96] and appears as an opportunity to display world leadership.[97]

Climate engineering expert agencies within the federal infrastructure

This internalisation of climate engineering into the federal bureaucracy brought a set of acronyms to the fore: NASA, NOAA, EPA, NCAR, NSF, but also DOE, USDA, DOD, DOS or USGS.[98] These acronyms stand for a group of federal agencies and departments, bundling climate engineering expertise within the state. Taken together, they bind the emerging politics of climate engineering to a historically grown expert infrastructure that has provided scientific expertise to US climate policy for many years.

As we will see in more detail in Chapter 4, the core of this group of expert organisations were essential in institutionalising the climate change issue in the federal bureaucracy. Roger Pielke Jr. has illustrated in detail how these agencies pushed a federally coordinated climate change agenda and 'developed expertise and responsibility for different aspects of the climate change issue' as soon as the late 1970s, first in the form of the National Climate Program, and then, since 1990, in the form of the United States Global Change Research Program (USGCRP).[99]

We will take a closer look at some of these agencies in Chapter 6 (see also Appendix). For now, it is simply relevant to note that this set of expert agencies embeds the recent rise of climate engineering as a controversial strategy of last resort in the broader trajectory of climate change expertise in US politics. This observation prepares one of this book's bigger themes, which is that the recent rise of climate engineering as a controversial 'Plan B' is yet another chapter in the federal cultivation of climate change expertise. To quote Roger Pielke Jr., these organisations make the emerging politics of climate engineering part of a story of 'how science was enlisted in support of policy development through the institutions of US government'. The status and role of these organisations reflects the reciprocal dynamic of observing and addressing societal challenges. The formulation of problem and response are necessarily intertwined.

This internalisation of climate engineering into the federal bureaucracy illustrates that political issues do not appear from thin air. Expert capacities, programs, and agencies are hardly created from scratch. Instead of amounting to a distinct, momentous decision, climate engineering arrived in the political sphere in this arena as a kind of climate policy introspective, a new category to assess, to inventory, and around which to further develop a historically grown federal infrastructure. This new category became meaningful by sorting and advancing what was already there, by taking stock of existing federal activities, which could then be adjusted, expanded and differentiated.

NOTES

1 This is how Alan Robock, a climatologist researching climate engineering, recently put the status of solar geoengineering (Robock (2020: 59)).
2 For the notion of 'matched struggles', see Baker (2017).
3 I focus here on the hearings, reports, and documents that appear in the studied corpus of US policy records on climate engineering (see Appendix for document corpus).
4 Keller (2009: 95). In a similar vein, Hilgartner has unpacked the 'drama' of science advice in his book, *Science on Stage*, where he describes the technical reports by the US National Academies as stylised productions (Hilgartner 2000).

5 Bentley Allan describes a similar process for the establishment of climate change as a governance object. He suggests that issues are *designated, translated*, and *problematised* as they take political shape and become objects of governance (Allan 2017).
6 See also Smallman (2020). Literature in Science and Technology Studies has illustrated the role of science in determining essential parameters of policy debates particularly in technical decision-making contexts (see, e.g., Pickersgill (2011); Hurlbut (2015)).
7 Keller (2009).
8 US House of Representatives, Committee on Science and Technology (2009: 222).
9 US House of Representatives, Committee on Science and Technology (2010b: 43).
10 US House of Representatives, Committee on Science and Technology (2009: 222).
11 Gupta and Möller (2019); see also Owen (2014).
12 Gupta and Möller (2019: 480).
13 See, e.g., Bellamy and others (2012) for a systematic assessment of various geoengineering appraisals.
14 US House of Representatives, Committee on Science and Technology (2009: 27f.).
15 Gupta and Möller (2019: 484).
16 See, particularly, US House of Representatives, Committee on Science and Technology (2009).
17 The initial 1992 report defines climate engineering primarily as an economical approach to tackling climate change, drawing on simple and largely favourable economic assessments and cost estimates, for which the report continues to be politically mobilised. See, e.g., (see, e.g., Lane in US House of Representatives, Committee on Government Reform (2006: 85); Schnare in US Senate 2007c: 109, 113). The 2015 volumes, in contrast, reflect the discursive shift that marked the political renaissance of climate engineering since the early 2000s and discusses climate engineering as a potential last resort for addressing increasingly dangerous impacts of climate change (US National Research Council (2015a, 2015b)).
18 US House of Representatives, Committee on Science and Technology (2009); US Government Accountability Office (2010a: 19; 2010b); Lattanzio and Barbour (2010); Bracmort and Lattanzio (2013); Bracmort, Lattanzio, and Barbour (2013). The IPCC reports AR 4 and AR5, too, refer to 'geo-engineering' (Intergovernmental Panel on Climate Change (2007, 2013)).
19 Royal Society (2009: ix). The House Science Committee defined geoengineering as 'the deliberate large-scale modification of the earth's climate systems for the purposes

of counteracting climate change' (US House of Representatives, Committee on Science and Technology (2009: 3)). See also the position statements of the American Geophysical Union (2009) and the American Meteorological Society (2013).
20 US House of Representatives, Committee on Science and Technology (2010b: IV).
21 US Government Accountability Office (2010a, 2010b, 2011).
22 US National Research Council (2015a: viii).
23 In 2019, for example, Representative Jerry McNerney introduced the Atmospheric Climate Intervention Act (H.R.5519, 2019); the 2020 Congressional Action Plan on Solving the Climate Crisis, too, refers to 'atmospheric climate intervention' measures; US House of Representatives, Select Committee on the Climate Crisis (2020: 525). See also US Global Change Research Program (2017) as the report refers to climate intervention (CI) measures.
24 See, e.g., Morrow (2014); Katz (2015); Baatz, Heyward, and Stelzer (2016); McLaren (2018); King (2019); Fiekowsky (2019).
25 US National Research Council (2015a, 2015b).
26 For an overview of these approaches, see, e.g., US National Research Council (2015a); Royal Society (2009).
27 Royal Society (2009: 6). This definition excludes measures which seek to capture CO_2 directly at the point of emission, such as carbon capture and storage, Carbon Sequestration from point sources, as 'clean coal', or in combination with bioenergy (BECCS), since the point of these measures is not to modify the climate, but rather to avoid further emissions to the atmosphere.
28 US National Research Council (2015a: 5; see also 67f.).
29 See, e.g., US National Research Council (2015a: 75f.).
30 US National Research Council (2015a: 56).
31 US National Research Council (2015a: 58).
32 For an overview see, e.g., US National Research Council (2015b); Royal Society (2009).
33 US National Research Council (2015a: viii). The National Research Council argued that albedo modification would be the 'physically more descriptive' term.
34 See, e.g., Latham and others (2012).
35 See, e.g., Pierce and others (2010).
36 Keith in US House of Representatives, Committee on Science and Technology (2009: 148).
37 Tollefson (2018).
38 See Caldeira in US House of Representatives, Committee on Science and Technology (2009: 23) for a taxonomy of the various relevant CDR approaches.
39 US National Research Council (2015a: 4).

40 See, e.g., Lane in US House of Representatives, Committee on Science and Technology (2009: 36; 31) and particularly Lackner's statement (168f.). Moniz and others (2019: 10); US House of Representatives, Select Committee on the Climate Crisis (2020).
41 US House of Representatives, Select Committee on the Climate Crisis (2020: 276).
42 US House of Representatives, Select Committee on the Climate Crisis (2020: 279).
43 US Government Accountability Office qtd. in US House of Representatives, Committee on Science and Technology (2009: 260). See also Royal Society (2009: 18). For a list of research and experiments in the area of ocean fertilisation, see also Oceanos (2018).
44 Lawrence and Crutzen in Blackstock and Low (2019: 90).
45 Lawrence and Crutzen in Blackstock and Low (2019: 90).
46 Keith in US House of Representatives, Committee on Science and Technology (2009: 148); see also US National Research Council (2015b: 4).
47 See, e.g., US National Research Council (2015b: 4).
48 This argument has its merit, of course, but it would also have to be applied to assessments of carbon dioxide removal, as neither of these approaches addresses human behaviour or emissions technologies as the anthropogenic causes of climate change.
49 US House of Representatives, Committee on Science and Technology (2009: 7, 26). See also Hale (2012); and Lin (2013). The literature has pointed out, however, that it is difficult to demonstrate such a causal mechanism between anticipated risk and behaviour, especially in the case of technological concepts which most people remain unfamiliar with so far.
50 See, e.g., US House of Representatives, Committee on Science and Technology (2009: 31, 149 f.).
51 See, particularly, Robock in US House of Representatives, Committee on Science and Technology (2009: 48f.).
52 Barrett in US House of Representatives, Committee on Science and Technology (2009: 315); see also Lane, Keith in US House of Representatives, Committee on Science and Technology (2009: 39, 150).
53 See, particularly, Rasch in US House of Representatives, Committee on Science and Technology (2009: 32, 39, 159, 161, 210 etc.); critical of this need, see Robock and Fleming in US House of Representatives, Committee on Science and Technology (2009: 45, 127). For an illustration of prerequisites of successful field studies, see for example, Robock in US House of Representatives, Committee on Science and Technology (2009: 50f., 119).

54 See, particularly, Rasch in US House of Representatives, Committee on Science and Technology (2009: 158f.).
55 See, e.g., Keith or Rasch in US House of Representatives, Committee on Science and Technology (2009: 149, 154).
56 Rasch in US House of Representatives, Committee on Science and Technology (2009: 157). See also, e.g., Robock, Rasch, Rusco, or Morgan in US House of Representatives, Committee on Science and Technology (2009: 45, 49, 119, 212, 252, 279).
57 US House of Representatives, Committee on Science and Technology (2009: 123).
58 See, e.g., US Senate, Committee on Environment and Public Works (2007: 14ff., 21, 47, 122, 150).
59 National Research Council (2015a: viii).
60 US National Academy of Sciences (2019, 2020); see also Intergovernmental Panel on Climate Change (2005, 2019).
61 US House of Representatives, Committee on Science and Technology (2009: 219f.); US Government Accountability Office (2010a, 2010b, 2011); US House of Representatives, Committee on Science and Technology (2010b); Bracmort and Lattanzio (2013); US House of Representatives, Select Committee on the Climate Crisis (2020).
62 In March 2010, the Bipartisan Policy Center (BPC), a think tank based in Washington D.C., for example, established a 'Task Force on Climate Remediation Research' to advise the US government on a strategy regarding climate engineering. In 2011, it presented its 'National Strategic Plan' (Long and others (2011)). In 2019 former Secretary of Energy, Ernest J. Moniz, presented a report via his Energy Futures Initiative which provided 'detailed implementation plans' for a federal program that would push commercial readiness of CDR measures (Moniz and others (2019: 1)). The report argued that a 'whole-of-government approach' would be necessary 'that reaches the mission responsibilities and research expertise' (Moniz and others (2019: 1)).
63 US House of Representatives, Committee on Science and Technology (2010b: V).
64 US House of Representatives, Committee on Science and Technology (2009: 222).
65 US Government Accountability Office (2010b: 4–5).
66 US Government Accountability Office (2010b: 5–6), emphasis added.
67 US Government Accountability Office (2010b).
68 US Government Accountability Office (2010b: 18).
69 See summary in US Government Accountability Office (2010b).

70 US House of Representatives, Committee on Science and Technology (2009: 220f.); Bracmort and Lattanzio (2013). The CRS presented its findings already in 2010, but it has provided an updated version of its assessment since then. In the UK, parallel efforts were undertaken and condensed into a report on 'The Regulation of Geoengineering', which was published by the Science Committee in March of 2010.
71 US House of Representatives, Committee on Science and Technology (2010b: 2).
72 See, particularly, the tabular overview in US Government Accountability Office (2010b: 31); and Bracmort and Lattanzio (2013: 29f.). For a general overview of relevant governance assessments and frameworks (2009–2015), see Morrow (2017).
73 See, e.g., US House of Representatives, Committee on Science and Technology (2009: 8, 21, 50, 112, 122, 130, etc.).
74 US Government Accountability Office (2010b: 26).
75 Bracmort and Lattanzio (2013: 22).
76 US House of Representatives, 111th Congress (2009: 33).
77 Gordon in US House of Representatives, US House of Representatives, Committee on Science and Technology (2010b: 79).
78 Lubchenco in US House of Representatives, Committee on Science and Technology (2010a: 79).
79 US House of Representatives, 112th Congress (2012: 28).
80 US House of Representatives, 112th Congress (2012: 28).
81 National Oceanic and Atmospheric Administration (2010: 3).
82 Baker (2017).
83 The alternative short title of the bill is America Creating Opportunities to Meaningfully Promote Excellence in Technology, Education, and Science Act.
84 *Public Law 110–69* (2007).
85 H.R.5519 (2019).
86 H.R.5519 (2019).
87 H.R.5519 (2019).
88 Aside from the Barrasso bills, there were, for example, the Energy and Water Development Appropriations Bill 2016 and the Consolidated Appropriations Act of 2016, which assigned funds to the Department of Energy to develop and commercialise direct air capture technology (US Senate, 114th Congress (2015: 82ff.); *Public Law 114–113* (2015)). In July 2019, Congressman Marc Veasey introduced the Fossil Energy Research and Development Act of 2019 which, if enacted, would direct the Department of Energy to establish 'one or more test centers… to provide unique testing capabilities for innovative direct air capture and storage technologies' (H.R.3607, 2019). And in January of 2020, the House Committee on Energy and Commerce presented a discussion draft for its CLEAN Future Act. The Act would

task the DOE with establishing 'a direct air capture technology prize program' for facilities which 'capture CO2 directly from the ambient air and capture more than 10,000 metric tons of CO2 annually' (US House of Representatives, Committee on Energy and Commerce (2020: 14)).

89 US House of Representatives, Select Committee on the Climate Crisis (2020: 281).

90 S.2744 (2009); S. 757 (2011); S.383 (2019). A number of additional documents simply list or discuss these suggested bills. See for example US Senate, 112th Congress (2011); US Senate, Committee on Energy and Natural Resources (2011); Wyden (2013).

91 S.2744 (2009: 2); S.757 (2011: 2).

92 S.383 (2019: 2).

93 S.383 (2019: 4).

94 'Algae carbon bio-capture is at a pilot-to-commercial stage at three coal-fired power plants in Australia. In the Land Down Under, they are advancing free enterprise deploying US technology. In the Land of the Manhattan Project and putting Men on the Moon, a prize short sells our proprietary knowledge. We as a Nation are better than this. Amending this bill should include financial and legislative support' (US Senate, Committee on Energy and Natural Resources (2011: 36)).

95 S.757 (2011: 3–4).

96 S.2744 (2009: 2); S.757 (2011: 2).

97 US House of Representatives, Select Committee on the Climate Crisis (2020: 278).

98 US House of Representatives, Committee on Science and Technology (2009: 5, 48, 54, 123, 172, 263ff.); US House of Representatives, Committee on Science and Technology (2010b: 28f.).

99 Pielke (2000a: 13). See tabular overview of the programs agency responsibilities in Pielke (2000a: 12).

CONCLUSION TO PART I

Around 2009, climate engineering gained policy traction as a 'bad idea whose time has come', as we have seen in Chapter 1.[1] Instead of invoking positive visions of socio-technical innovation, it emerged as a controversial last resort option, a daunting Plan B that humanity must consider when facing the impending climate crisis. This conflicted status of climate engineering raises important questions for science and science policy scholarship: depending on where we stand, the question is either how these measures have earned a spot in policy agendas despite their enormous scientific complexities and fierce political contestation. Or the question is why these measures have only emerged as a last resort measure despite their much-reiterated promise to tackle one of the most pressing challenges of our time. By raising these questions, this book seeks to dissect this contested innovation process and to confront notions of choicelessness in dealing with this debate. Specifically, it retraces the 'career' of climate engineering as a product of historically grown science-politics interrelations. This book asks for the kinds of science-politics alliances that came to cast climate change as an engineering challenge and established the concept of technological climate intervention as the controversial policy measure that it is today.

In the first part of the book, we began this journey through the lens of politics. We saw how climate engineering began to materialise in the political realm in two distinct contexts. First, Chapter 2 illustrated how climate engineering became established as a political issue in its own right when experts and policymakers began producing an 'official record' on the topic around 2009. Climate engineering materialised here as subject to a kind of 'staged advice'. Via congressional expert testimonies, legislative and scientific assessment reports, scientific experts essentially 'assembled' climate engineering as a potential policy measure.[2] These records established climate engineering as matter of fact(s); they defined its status as a potential policy tool. As a result, these records essentially seem to de-politicise the emerging politics of climate engineering. Despite their political importance in determining the very stakes of the debate, climate engineering appears here in neat packages of relevant facts. These are apparently shielded from the future politics on this issue that merely loom over these accounts in the form of informed decisions that will follow somewhere down the line.

Secondly, Chapter 2 suggested how climate engineering became structurally internalised by the political system. Drawing on the 'official record' on the issue, climate engineering appeared as a category along which policymakers and scientific advisors 'inventoried' already existing expert capacities within and beyond the federal bureaucracy and respectively sought to steer their further development. Climate engineering in this context appeared as both already existing and entirely new: it provided a category that shed new light on a historically developed infrastructure of climate science expertise in the federal bureaucracy.

Making sense of the emerging politics of climate engineering is thus not merely a question of who or what managed to place a somehow predetermined issue on the political agenda. It is rather a question of how politics came to look at the issue of climate change in these terms of climatological intervention and control; how the state adopted this perspective; how it both cultivated and internalised this particular mode of observing, problematising, and tackling climate change. To quote Allan, making sense of the emerging politics of climate engineering requires considering how 'the history of the governance object is bound up with the history of knowledge production in scientific disciplines and expert groups'.[3] This first part of the book has set the stage for this endeavour, hinting at the complex interplay of science and politics in shaping this career of climate engineering. It has suggested how science comes to bear on politics, not primarily in an advisory role or as a robust and solid ground for political decision-making. Again, the politics of climate engineering cannot be merely boiled down to a discrete decision which needs expert guidance. Much more importantly, we have seen how the very notion of climate engineering essentially links various lines of scientific research to the political challenge of governing climate change. Climate engineering, in this sense, is neither simply a line of scientific research that has become politically relevant, nor is it simply a political project of control that has guided scientific research. Instead, climate engineering emerged precisely from the interrelation of science and politics. This suggests that in order to make sense of climate engineering as something new and controversial, we need to understand it as something historically contingent. That is, to reflect on how to move forward, we first must look back.

NOTES

1 Kintisch.
2 See also Allan (2017).
3 Allan (2017: 139).

PART II

EARLY VISIONS OF CONTROL

'Modification plans always were couched in the context of the pressing issues ... of their eras... Each generation ... has had its own leading issues for investing in technologies of control'.
Fleming (2010: 265)

3

WHERE DOES THE STORY BEGIN?

AS I SUGGESTED AT THE END OF THE LAST CHAPTER, IN ORDER TO MAKE sense of where we stand now, we first have to look back. So, where does this story begin? In this second part of the book, we will see how trying to determine the historical roots of what today is called climate engineering forces us to embark on a turbulent journey through the history of climate science to the turn of the twentieth century (and sometimes earlier to the seventeenth and eighteenth century) when human efforts to modify climatic conditions gradually emerged as a critical political project in North America and Europe.[1] Exploring these historical perspectives does more than add mere context or background to the story of this book. They are essential to understanding the recent rise and status of climate engineering. On the one hand, they unpack the roots of ideas about 'engineering' or deliberately modifying the climate. These perspectives explain how experts came to look at global warming as an engineering issue, an issue that might be addressed by targeted techno-scientific intervention and control. On the other hand, they suggest why climate engineering did not, however, emerge as a 'Plan A' in the face of global warming, but instead gained traction as a daunting possibility, a 'bad idea whose time has come'. Diving into the longer history of climate engineering thus explains how we arrived at the present and provides the grounds for staging a meaningful debate over how to move forward.

In this chapter, we will see how visions of control, deliberate intervention, modification, or 'engineering', came to define relations between a nascent field of climate science and politics during the first half of the twentieth century. These visions of control 'matched'[2] the scientific to political struggles of the time. They fostered political interest in climate expertise and were instrumental in establishing the institutional and material infrastructure that modern climate science now rests on.[3] This chapter will provide a brief glimpse into the longer-standing history of how experts came to look at climatic change in a way that would suggest the option and potential of its targeted 'engineering'. It traces how this particular gaze onto the issue was assembled, providing the historical roots for later notions of climate engineering.

Before we embark on this journey, it should be noted that this effort necessarily implies working with a moving target. I do not wish to suggest that these early visions of climate modification and control – indeed the very concept of what was understood as 'climate' – has been the same for the past hundred or so years. Climate science, as we will see, was part of meteorology for most of the twentieth century, and that means that 'climate and weather were not just intimately connected, they were essentially identical'.[4] It was not really until the 1980s that the climate emerged as a global category, understood as more than aggregated weather phenomena. But this essential interrelation between scientific insights and concepts on the one hand, and the social order on the other, is precisely part of this story. The goal of the following pages in this sense is to follow the experts' accounts through their respective historical settings and to understand how these accounts have incrementally assembled a vision of deliberate climate modification and control that suggests climate engineering as a potential remedy against global warming. So, when I speak of 'climate' in the following, then this necessarily comprises a very different scientific concept than the one that emerged during the second half of the twentieth century.

THE DISCOVERY OF A 'GRAND POSSIBILITY': NOTIONS OF CLIMATE MODIFICATION BEFORE THE MID-TWENTIETH CENTURY

On 14 August 1912, an Australian newspaper featured some promising news on the prospects of climatic changes, or, more specifically, on the 'considerable' impact that carbon dioxide emissions may have on the Earth's temperature:

The furnaces of the world are now burning about 2,000,000,000 tons of coal a year. When this is burned, uniting with oxygen, it adds about 7,000,000,000 tons of carbon dioxide to the atmosphere yearly. This tends to make the air *a more effective blanket for the earth* and to raise its temperature. The effect may be considerable in a few centuries.[5]

This newspaper snippet illustrates how initial observations of what today is considered one of the greatest challenges to humankind, not only motivated optimistic reactions, but seemed to promise great potential. Scientists viewed carbon emissions not as a potential problem, but as a potential solution: 'warming seemed a good thing'.[6] In fact, early observers of human impacts on the climate sensed a 'grand possibility' to deliberately regulate the 'future climate of the earth'.[7] In what follows, we will trace how human impacts on the climate incrementally emerged as an object of scientific observation and political interest during the turn of the twentieth century and how this fuzzy picture of climatic change fostered initial hopes of deliberate modification.

Observing, charting, mapping: The rise of weather stations

The emergence of a global system of meteorological data collection spanned several centuries and varied substantially across different nations.[8] In the United States, coordinated data collection emerged from scattered individual efforts, dating all the way back to the seventeenth century. In his 'Short Bibliography of United States Climatology' from 1918, Harvard Professor Robert Ward refers to Rev. John Campanius, 'who, in 1644–45, at the Swedes' Fort, near Wilmington, Del. kept what is believed to have been the first regular record of the weather on the North American continent'.[9] These efforts gained political support through several of the so-called founding fathers who began charting weather and water temperatures during the eighteenth century.[10]

Early notions of climate modification appear here already in the context of American settler colonialism. Historian Jim Fleming recounts how the idea that the North American climate could be improved by cultivating the land was a critical theme in colonial America. 'Colonial promoters' suggested that by settling the land – clearing the forests and cultivating the soil – the climate would become more moderate and pleasant.[11] This narrative became an integral part of the colonial project. Proponents envisioned that the so improved and cultivated climate would provide a 'proper nursery of genius, learning, industry and the liberal arts'.[12] Critical voices warned of the adverse effects of these early climate modification schemes, suggesting that deforestation was in fact leading to desertification and crop failure.[13]

The nineteenth century brought coordination and cooperation to this new endeavour, as interested parties attempted to standardise and broaden the scope of meteorological observations. Most of the early efforts to coordinate data collection were state funded and driven by military and regional agricultural interests.[14] The first nationally coordinated data collection effort in the United States was initiated by an 1816 order from the Army's Medical Department, which made it obligatory for every surgeon to keep a weather diary.[15] This military order generated a detailed record of weather data.[16] In 1849, the Smithsonian Institution initiated a 'fairly extended system of observations' employing as many as three hundred and fifty observers. This work was later transferred to the War Department (as were all of the Smithsonian's meteorological initiatives). Finally, in 1870, the US government established the National Weather Service (NWS). As with previous initiatives, the work of the Weather Service was closely linked to the military. Military posts served as central meteorological observation points and the army's Chief Signal Officer oversaw the most important tasks being conducted by the newly created institution.[17] Jim Fleming suggests that the Weather Service essentially served as a 'national surveillance force'. Weather patterns were just one of the 'threats to the domestic order' that were observed here, from 'striking railroad workers, Indian uprisings on the frontier, locust outbreaks, and natural hazards to transportation, commerce, and agriculture'.[18]

During these early years, 'climate research' in the United States thus primarily consisted of a network of amateur observers, charting and mapping weather patterns and temporal variations through an infrastructure of state and private weather services. These weather services largely amounted to 'ad hoc efforts staffed by volunteers' and were strongly tied to practical – agricultural, security – interests.¹⁹ The relevant theoretical insights that would contribute to scientific theories of climatic change happened elsewhere.

Scientific advances and the emergence of the hothouse theory

Progress in physical theories of global climate change was not systematically tied to this emerging infrastructure of practically oriented weather services.²⁰ Instead, it was generated as part of a series of individual efforts without a clear institutional centre, mostly scattered across Europe.²¹ Only in hindsight can these efforts be understood as contributing to a joint trajectory. At the time, it rather seemed like a big mess, 'each expert championed a personal theory about *the* cause of climate change' as Spencer Weart suggests.²²

There are many rich histories of this 'discovery' of global warming and the greenhouse effect.²³ What follows can hardly serve as a comprehensive summary of these important insights. I merely want to focus on some of the central insights and relevant theories as they relate to the question at hand, namely how efforts to understand were connected to efforts to deliberately modify and control the climate.

In the 1850s and 60s, Irishman John Tyndall, a professor of natural philosophy at the Royal Institute of Great Britain, experimented with the radiative potential of various gases in the atmosphere, including carbon dioxide, ozone, and water vapour.²⁴ By 1861, Tyndall had concluded that these gases were responsible for 'all the mutations of climate which the research of geologists reveal. [...] They constitute true causes, the extent alone of the operation remaining doubtful'.²⁵ Tyndall eventually demonstrated that trace atmospheric constituents effectively retained heat radiation, attesting to what was then called the 'hot-house theory'.²⁶ Building on this research, as well as the work of the French physicist Joseph Fourier, and others, Swedish physicist Svante Arrhenius and meteorologist Nils Gustav Ekholm studied the impact of carbon dioxide on the Earth's temperature.²⁷ Earlier scientific findings on the connection between carbon dioxide and the climate had focused primarily on geophysical variations in the carbon cycle due to volcanic eruptions, vegetation, and other natural factors.²⁸ Arrhenius extended these insights by developing a model that could predict the onset of Ice Ages[29] – an issue that motivated much climate research at the end of the nineteenth century.[30]

Although scientists initially believed that anthropogenic carbon emissions played a rather negligible role in climate processes, this understanding shifted dramatically shortly thereafter.[31] About a decade after Arrhenius had presented his prominent paper, 'On the Influence of Carbonic Acid in the Air upon the Temperature on the Ground',[32] to the Stockholm Physical Society, he began to recognise the 'noticeable degree' to which the percentage of atmospheric carbon dioxide had changed in response to 'the advances of industry […] in the course of a few centuries'.[33] By slowly detecting the societal impact on the geologic processes of the climate system, Arrhenius and Ekholm incrementally united human agency and the natural climate system at the dawn of the twentieth century.[34]

This early theoretical advancement of climatological knowledge, however, did not include prophetic concerns about a global warming trend.[35] In fact, the initial problematisation of climatic change inverted today's problematisation of the issue: well into the twentieth century, the assumption was that humankind would soon experience a new Ice Age.[36] Emerging insights into the possible human impact on the climate via the burning of fossil fuels thus sparked hope, namely, the hope of being able to avert the daunting crisis of a freezing planet.[37]

In effect, these insights prompted early visions of control, suggesting the potential to deliberately modify the climate. Writing in 1901, Ekholm was amazed by the sheer potential and prospect of human impacts on the climate. His observations seemed to promise the possibility

> […] *that Man [sic] will be able efficaciously to regulate the future climate of the earth and consequently prevent the arrival of a new Ice Age.* […]. It is too early to judge of how far Man might be capable of thus regulating the future climate. But already the view of such a possibility seems to me so grand that I cannot help thinking that it will afford to Mankind hitherto unforeseen means of evolution.[38]

Fleming therefore described Ekholm as an 'early and eager spokesman for anthropogenic climate control'.[39] Observations such as the one by Ekholm or the Australian newspaper from the beginning of this chapter, illustrate just how differently climatic change and deliberate climate intervention were related in this historical setting compared to today. The prospect of anthropogenic climate change appears here as an 'unforeseen means of evolution', hardly comparable to the daunting prospect of dangerous global warming that we are confronted with today. This begins to suggest just how deeply embedded scientific insights are in their respective historical times.

We must fast forward several decades from Ekholm's observations to see how empirical evidence of a warming trend that could be confirmed by detailed meteorological records pushed the carbon dioxide theory of climatic change back into the scientific limelight.[40] In 1938, Guy Stewart Callendar, a British engineer, re-established the scientific validity of the carbon dioxide theory of climate change[41] when he presented a data set that clearly demonstrated a warming trend. Although he was not a professionally trained meteorologist, he 'had the audacity to stand before the Royal Meteorological Society in London'[42] and formulate the carbon dioxide theory in its 'recognizably modern form'.[43] To be sure, weather researchers continued to view anthropogenic warming as unproblematic. In the context of continued fear of the coming Ice Age, Callendar, too, believed that his findings ensured the indefinite delay of the 'return of the deadly glaciers'.[44]

At the turn of the twentieth century, the relationship between scientific progress in climatology (as a subfield of meteorology) and the state was thus defined by a notable divide. On the one hand, there was scientific progress and disciplinary advancements scattered across many individual projects of scientific curiosity, and on the other, there was the dominant system of state-supported weather services that employed most 'meteorologists' of the time. This divide was particularly substantial in the United States, as we have seen, where climate research was mostly restricted to the Weather Bureau and strongly tied to military and practical agricultural interests.[45] Through the first half of the twentieth century, most professionals at the Bureau 'lacked any college degree'.[46] Indeed, the first meteorological university department in the United States was not established until 1928 (at the Massachusetts Institute of Technology).[47] According to

Spencer Weart, the predominant assumption during this time was that 'a canny amateur with no academic credentials could predict rain as successfully as a Ph.D. meteorologist'.[48] The divide between academic research and pragmatic weather services was somewhat less dramatic in other countries. In Europe, for example, meteorology was considered equally prestigious as astronomy and theoretical research in the field was conducted in academic institutions in Norway, Sweden, England, and Germany.[49] But 'everywhere', as Edwards suggests, it was national weather services and not universities or other academic institutions that employed the vast majority of meteorologists. Climate research, in other words, mainly boiled down to charting and forecasting the weather during the time and was seen 'as a form of practical work, rather than a research science'.[50] There was no political interest in advancing scientific theoretical work in the field.

THE GEOPOLITICAL CHALLENGES OF THE TWENTIETH CENTURY: CALCULATING, PREDICTING, AND CONTROLLING THE CLIMATE

The beginning of World War II changed this outlook substantially. Political (and especially military) interest in expanding techno-scientific control over climatic conditions 'matched' scientific interests. This lead to the setup of a massive infrastructure of climatological expertise and effectively advanced the emergence of climatology as a bounded discipline.[51] As we will see later, this infrastructure eventually 'discovered' anthropogenic climate change as an issue of environmental safeguarding, effectively questioning the hopes of techno-scientific intervention that had driven the very setup of this infrastructure. Put differently, the infrastructure that was devised to make the Earth's climate politically legible and controllable would later put the very prospect of control into question.

Linking political and scientific agendas around climatological expertise

The geopolitical challenges of the mid-twentieth century transformed meteorology and oceanography, into what Zeke Baker refers to as a political 'high-stakes issue'.⁵² Baker recounts how in the United States especially, climate research became a well-established scientific discipline in the face of these challenges. The urgent military need for progress in meteorological and oceanographic expertise pushed scientific boundaries, uniting the formerly disparate state-funded and scientific efforts in climate research, and – in Baker's terms – directly 'matched' political and scientific challenges.

World War II generated a boom in both meteorological personnel and observational infrastructure. At the beginning of the war, US meteorologists remained institutionally bound to the Weather Bureau and isolated from academic networks. As indicated earlier, this marked a stark contrast from Europe. Thus, when major US universities began developing academic programs in the early 1940s, almost all were directed by Scandinavians who taught Bergen school theories and methods.⁵³ The so-called Bergen school, named after their Norwegian location, was a group of researchers linked to the Norwegian meteorologist Vilhelm Bjerknes who 'redefined the basic concepts of weather prediction' and provided the cornerstone for scientific meteorology.⁵⁴ Bjerknes essentially conceptualised meteorology as an exact atmospheric science, providing the theoretical foundations for climate modelling. The atmosphere became understood in this context as a 'purely mechanical and physical phenomenon', an 'air mass circulation engine', as Gabriele Gramelsberger explains. This engine is driven 'by solar radiation and gravitational forces expressed in local differences of velocity, density, air pressure, temperature, and humidity'.⁵⁵ With this understanding, Bjerknes outlined the basic equations for a General Circulation Model (GCM) of the climate. The problem, however, was that they were too complex to solve with the then existing analytical methods. It would take the advent of computers to turn these equations into action and run the first numerical climate models, as we will see in a bit.

The Bergen school scientists 'worked diligently' to make their insights known internationally.⁵⁶ In the end, Fleming suggests that it was a graduate student, Anne Louise Beck, who drew the attention of the US Weather Bureau to the Bergen school methods for the very first time. After a year of working alongside Bjerknes, Beck published her thesis in the *Monthly Weather Review* making it visible across the Atlantic. In addition, the military challenges of an air-based war directly matched the scientific challenges of understanding, observing and bringing new perspectives to atmospheric circulation dynamics.[57] As a result, the war facilitated the development of critical observational and computational infrastructure, including radar, satellites, and the electronic computer.[58] In a matter of years, tens of thousands of Americans were trained as meteorologists and technicians. Some observers of the era have estimated a 1500% growth in professional personnel during the war years.[59] Established only eight years earlier, the army's Air Weather Service (AWS), for example, employed 19,000 people by 1945.[60] This military connection persisted as we will see. Long after the war had ended, most American meteorologists were still linked to the military.[61]

By the end of the war, basic scientific advancement in climate science had become a national strategy as part of the escalating conflict with the Soviet Union. Within this context, climate science emerged as a bounded scientific field based on its capacity to 'shape the national security state',[62] and it thrived because it was directly linked to the continuing geopolitical challenges of the time. This distinct science-politics configuration facilitated substantial advances in the problematisation of both climate change and climate control.

The International Geophysical Year (IGY), a set of projects spanning 1957 and 1958, as well as the first computer model experiments, were both essential developments in the problematisation of climate change as a societal challenge. The IGY was a response to the increasingly intolerable disciplinary fragmentation of climate-related research in the middle of the twentieth century. Researchers from a plethora of distinct, highly specialised subfields were struggling to collaborate across disciplinary and geographical borders. The IGY addressed this challenge by invoking political hopes for military applications of the resulting knowledge:[63] '[N]ational security and scientific internationalism coalesced around a broad program of rational mastery [...]'.[64] The resulting project offered opportunities for international scientists from different disciplines to work together on 'interdisciplinary research projects grander than any attempted before'.[65] As part of the IGY projects, Charles David Keeling commenced measurements of carbon dioxide (CO_2) on top of the active Mauna Loa volcano in Hawaii.[66] Keeling's measurements resulted in the infamous Keeling Curve, an 'icon' of anthropogenic climate change today.[67] His insights were able 'put a capstone' on the earlier work by Tyndall, Arrhenius, Callendar, and others.[68]

The 1950s and 1960s also witnessed the computation of the earliest climate models. As we saw earlier, Vilhelm Bjerknes and Felix Exner had identified the basic equations of atmospheric dynamics in the early twentieth century. Starting in the 1940s, an increasing focus on numerical weather prediction linked scientific challenges related to the representation of atmospheric dynamics to military strategic challenges, such as aviation safety.[69] The subsequent advent of the first calculating machines – and eventually the electronic computer – meant that Bjerknes and Exner's equations could be mathematically solved and computed.[70] What follows is a rush of events, driven both by US American and Scandinavian scientists.[71] In the US, Jule Charney and others successfully computed several prognoses on the Electronic Numerical Integrator and Computer (ENIAC).[72] Gabriele Gramelsberger and Johann Feichter recount how, beginning already in 1948, Charney and his colleagues 'developed the very first computer model for weather forecasting, a simple barotropic model with geostrophic wind for the area of the United States of America'.[73] In 1950, the scientists then ran 'the first numerical experiment ever conducted in meteorology'.[74] And in the spring of 1953, Swedish scientists generated what would be the very first real-time weather prediction on the Swedish Binary Electronic Sequence Calculator (BESK).[75] The scientists were successful at 'beating the actual weather by some ninety minutes'.[76] These developments incrementally transformed meteorology from a strictly descriptive field of rather practical ambitions into an increasingly theory- and model-based science.[77]

The rise of deliberate climate modification ...

After the war ended, the boom in numerical weather prediction was sustained by the growing political interest in modifying atmospheric processes.[78] Global climate models provided a critical link between the advancement of political and scientific infrastructures, securing financial and organisational resources for both sides. The newly established infrastructure of meteorological expertise enabled observations at higher resolutions and thus generated visions of prediction and control in relation to the climate.

The work of Harry Wexler, one of the most renowned meteorologists of the mid-twentieth century, is a case in point.[79] In Wexler's career, many of the institutions, concepts, and infrastructures that had been shaping the establishment of climate expertise at the interface of science and politics coalesced. Wexler became an internationally recognised figure pushing the boundaries of new technologies. He pioneered the use of electronic computers for climate modelling purposes and was centrally involved in the management of US atmospheric observations via satellites and rockets. Wexler worked for the US Weather Bureau throughout his career and in 1961 was appointed to represent the United States in negotiations with the Soviet Union on the joint use of meteorological satellites.

In the 1950s, Wexler began to dedicate Bureau resources to the critical exploration of climate control.[80] Somewhat ironically, Wexler's appraisal of climate intervention options was an important step in the further advancement of this research. As a well-established and renowned scientist, Wexler was able to supply this line of inquiry with professional legitimacy, despite his critical attitude towards these measures.[81] This ambiguous engagement with the outlook of climate control is thus notable as it parallels the stance of much of the scientific exploration of climate engineering today. In other words, Wexler's pioneering engagement with climate control seems driven by a somewhat sceptical scientific curiosity, a curiosity that, on the one hand, effectively advanced the debate, drawing attention to not only the possibility and outlook of such an endeavour, but also to some of the technical intricacies. Meanwhile, on the other hand, it was guided by concern and distress for what this might imply.

In 1958, Wexler criticised the 'tempting' possibility of modifying basic atmospheric radiation, particularly by altering the Earth's reflectivity.[82] Wexler concluded that

> when serious proposals for large-scale weather modification are advanced, as they inevitably will be, the full resources of general-circulation knowledge and computational meteorology must be brought to bear in predicting the

results so as to avoid the unhappy situation of the cure being worse than the ailment.[83]

Four years later, he pursued his sceptical fascination with anthropogenic climate control in a lecture entitled, 'On the Possibilities of Climate Control', which was addressed to technical audiences across the country.[84] Wexler was explicitly concerned here not with the prospect of local weather modification schemes – such as precipitation control or 'rain-making' – but with planetary-scale modifications that would result in 'large-scale effects on general circulation patterns in short or longer periods, even approaching that of climatic change'.[85] Notably, this concern included both deliberate as well as unwanted, inadvertent modifications of the climate, a distinction which emerged as essential during these years as we will see later.

In his lectures, Wexler presented a 'purely hypothetical' taxonomy of different kinds of climate modification schemes which 'man might attempt deliberately to exert and also which he may now be performing or will soon be performing in ignorance of its consequences'.[86] This taxonomy of deliberate as well as inadvertent climate modification schemes included approaches to both heat and cool the planet. He suggested (a) increasing the global temperature 'by injecting a cloud of ice crystals into the polar atmosphere by detonating 10 H-bombs on the Arctic sea ice; (b) decreasing the global temperature by installing a global sun shield, by launching 'a ring of dust particles into equatorial orbit to shade the Earth'; (c) suggesting that we 'warm the lower atmosphere and cool the stratosphere by injecting ice, water, or other substances into space'; and finally (d) 'destroy[ing] all stratospheric ozone, raise the tropopause, and cool the stratosphere by up to 80°C [144 °F] by an injection of catalytic de-ozonizer such as chlorine or bromine'.[87]

In that same year, 1962, after witnessing how substantial progress in climatological research further fuelled scientists' interest in climate modification, Wexler warned a United Nations panel of the 'inherent risks' of this endeavour.[88] Only a few years later, however, many national research organisations and scientific advisory bodies, such as the National Science Foundation (NSF), the President's Science Advisory Committee (PSAC), the RAND Corporation, and the National Academy of Sciences (NAS), began to embrace the idea that increasingly better understanding of climatological processes would eventually lead to the option of their deliberate modification and 'engineering'.[89]

In some respects, the debate that emerged over climate control during these years is thus strikingly comparable to discussions over climate engineering today.[90] David Keith suggests that 'the case for continuity' between the emerging interest in deliberate climate modification during the 1960s and the debate over climate engineering today primarily 'rests on the similarity of proposed technical methods, the continuity of citations to earlier work', as well as 'a similarity of debate about legal and political problems' with the discussed measures.[91] The technical concepts that were being discussed during the 1960s bear 'a strong similarity' in particular to what is now labelled as solar radiation management.[92] Not only Wexler's work, but also many of the above-mentioned assessments of climate control explored the potential of modifying the Earth's reflectivity (albedo) in order to bring about deliberate changes in the global temperature. Technologies to remove CO_2 from the atmosphere, in contrast, were not yet discussed, despite the fact that carbon dioxide emissions were increasingly recognised as a problem, leading to 'inadvertent modification of atmospheric processes'.[93]

A relevant difference between the debate in the 1960s and today concerns the motivation behind efforts to deliberately modify the climate. Keith suggests that during the 1960s, the focus rested on an 'improvement of the natural state or mitigation of natural hazards, whereas the aim of recent geoengineering proposals is the mitigation of anthropogenic hazards'.[94] The problematisation of deliberate climate modification as a potential political tool, as a response measure, or device of societal transformation, in other words, must be understood in the particular context of its time. Anthropogenic climate change as the defining reference problem for the current debate over climate engineering was only beginning to appear as an issue of global political significance during these years.

... And the rise of inadvertent climate modification

The rising interest in the potential of climate modification during the 1950s and 1960s not only pushed hopes for techno-scientific control over the atmosphere further into the science-policy limelight, but it also foreshadowed an imminent turning point in debates about such measures. Towards the latter half of the 1960s, environmental concerns appeared as increasingly prominent reference points in explorations of climate modification. Assessments of a grand possibility incrementally converged with the incremental discovery of a potential problem. Military interests had dominated efforts in weather control, particularly precipitation control during the Cold War. Cloud seeding discoveries at the General Electric Corporation after the war, for example, significantly advanced both commercial and military interest in these options.[95] During the 1960s, scores of deliberate weather and climate modification programs operated in the United States[96], as well as in the Soviet Union. Climate modification efforts continued to be largely motivated by the depiction of technological power as central to national strategic interests.[97]

Towards the end of the 1960s, this military exploration of climate modification was punctuated by the recognition of unwanted side effects. For the first time, 'inadvertent climate modification' started to appear relatively consistently as a topic. This 'inadvertent modification' was not yet problematised as a societal issue with global political consequences, but rather as an involuntary consequence of 'the technological evolution of man'.[98] Assessment bodies, such as the National Academies remained largely enthusiastic about the 'exciting' prospect of 'man's ... power to modify his atmospheric environment'. Yet, incrementally, this power now seemed to provide both 'challenge and opportunity:'

> The challenge and opportunity presented to the world by the prospect of man's achieving the power to modify his atmospheric environment is one of the most exciting of the long-range aspects of the subject. We are dealing with the possible consequences of a new and perhaps enormous power to influence the conditions of human life. Its potentialities for beneficial application are vast.[99]

In these expert observations, climatological research not only promised a great prospect – namely to deliberately modify the global climate – but it also began to foster a cautious sense of humility in the face of this 'enormous power'.[100] What would a couple of years later be discussed as problem and response, emerged here as two sides of a single equation: deliberate or 'conscious' modification was juxtaposed with 'inadvertent human intervention'.[101]

This chapter has provided just a brief glimpse into the entangled histories of climate science and climate modification schemes. While scientific research and state interests have initially evolved separately in the United States, the geopolitical challenges of early and mid-twentieth century forged strong alliances between the two. Notions of modification and control were critical drivers of these alliances and thus figured prominently in the material and institutional establishment of climatology during this time. A dive into the early history of climate science has suggested that before human impacts on the climate were considered as a problem of global societal significance, they were seen as a major new opportunity, an exciting potential of the 'enormous power of man'.

From this perspective, we might thus expect that the impending politicisation of anthropogenic climate change during the 1980s would lend renewed political attention to deliberate modification schemes. We might expect, in other words, that what is referred to today as 'climate engineering' would gain unprecedented steam in the face of the discovery of this new 'grand societal challenge'. As we will see in the next chapter, however, the years that followed – especially the late 1960s to the 1990s – would tell a different story.

NOTES

1 See, e.g., Fleming (1998); Locher and Fressoz (2012); Baker (2017).
2 Baker (2017).
3 See, particularly, Edwards (2006, 2010).
4 Miller (2004: 51); see also Edwards (2010).
5 *Rodney and Otamatea Times, Waitemata and Kaipara Gazette* (1912, emphasis added).
6 Weart (2008: 7).
7 Ekholm (1901: 61, emphasis added).
8 Fleming (1998: 34–41).
9 Ward (1918: 179).
10 Ward, for example, pointed to Jefferson, Madison, and Franklin (Ward (1918: 180)).

11 Fleming (1998: 32).
12 Hugh Williamson 1811, qtd. in Fleming (1998: 24).
13 See, e.g., Fleming (1998: 21f.).
14 Baker (2017: 7); Fleming (1998: 34).
15 Ward (1918: 180ff.).
16 In 1890, the data was transferred to the Weather Bureau (Ward (1918: 180)).
17 Ward (1918: 181).
18 Fleming (2010: 168).
19 Fleming (1998: 34).
20 See, e.g., Fleming (1998: 55f.); Weart (2008); Edwards (2010: 72f.).
21 Edwards (2006: 231); Baker (2017: 6).
22 Weart (2008: 18).
23 See particularly Fleming (1998); Weart (2008).
24 See, e.g., Fleming (1998: 66).
25 Edwards (2010: 73).
26 Arrhenius (1896: 237).
27 See, particularly, Arrhenius (1896: 237); Ekholm (1901). For historical accounts of relevance, see Fleming (2010: 4); Fleming (1998: 65f.); Edwards (2010: 72f.); Weart (2008: 5f.). According to Fleming, 'simple claims about the discovery of the greenhouse effect are impossible to sustain' (Fleming (1998: 80)).
28 Arrhenius (1896: 272).
29 Fleming (1998: 79).
30 Gramelsberger and Feichter (2011: 16).
31 Fleming (1998: 78).
32 Arrhenius (1896). See also Fleming (1998: 76f.).
33 Arrhenius (1908: 54).
34 Fleming (2010: 4).
35 See, e.g., Fleming (1998: 82); Weart (2008: 7).
36 This hypothesis was established by James Croll or Lord Kelvin; see, e.g., Fleming (2010: 5); Weart (2008: 16).
37 See, e.g., Gramelsberger and Feichter (2011: 16).
38 Ekholm (1901: 61, emphasis added).
39 Fleming (1998: 111).
40 Weart (2008: 1).
41 Edwards (2010: 76f.); Fleming (2010: 5).
42 Weart (2008: 2).
43 Fleming (2010: 5) or Gramelsberger and Feichter (2011: 16).
44 Callendar qtd. in Gramelsberger and Feichter (2011: 16).
45 Baker (2017: 6).

46 Weart (2008: 11); see also Edwards (2010: 90f.).
47 Baker (2017: 7).
48 Weart (2008: 11).
49 Harper (2008: 12); see also Baker (2017: 7).
50 Edwards (2010: 90).
51 For the notion of a 'knowledge infrastructure' in this context, see particularly Edwards (2010).
52 Baker (2017: 8).
53 Baker (2017: 8).
54 Edwards (2010: 90f.). For the relevance and work of Bjerknes, see also Gramelsberger (2011); Fleming (2016: 13–75).
55 Gramelsberger (2011: 297).
56 Fleming (2016: 9).
57 Baker (2017: 8).
58 Fleming (2010: 169).
59 Baker (2017: 8).
60 Fleming (2010: 169).
61 See, e.g., Fleming (2010: 169); Baker (2017: 17).
62 Baker (2017: 14).
63 Weart (2008: 33).
64 Baker (2017: 16).
65 Weart (2008: 33).
66 Keeling (1998: 35f.); Gramelsberger and Feichter (2011: 16); Weart (2008: 34).
67 Gramelsberger and Feichter (2011: 1).
68 Weart (2008: 37).
69 See, e.g., Baker (2017: 10); Fleming (2010: 110).
70 Fleming (2010: 192); Gramelsberger (2011: 297); see also Fleming (2016).
71 For detailed accounts of the origins and trajectory of numerical weather prediction (NWP), see, particularly, Harper (2008); Edwards (2010: 111–37); Gramelsberger (2011); Gramelsberger and Feichter (2011); Fleming (2016).
72 Gramelsberger (2011: 297).
73 Gramelsberger and Feichter (2011: 31–32).
74 Gramelsberger (2011: 297).
75 Fleming (2016: 157); see also Edwards (2010: 129).
76 Fleming (2016: 157).
77 Gramelsberger and Feichter (2011: 4, 20).
78 Baker (2017: 11, 17).
79 Fleming (2010: 214–16); see also Fleming (2016: 179–84).
80 Baker (2017: 18).

81 Fleming (2010: 214).
82 Wexler (1958: 1059).
83 Wexler (1958: 1065).
84 Fleming (2010: 216).
85 Wexler qtd. in Fleming (2010: 217).
86 Wexler qtd. in Fleming (2010: 217–18).
87 Wexler qtd. in Fleming (2010: 218).
88 Fleming (2010: 6).
89 President's Science Advisory Committee (1965); US National Research Council (1965); RAND Corporation (1969); see also US National Academy of Sciences (1973). Notably, the NAS assessments were conducted during the war in Vietnam, in which the United States utilised weather modification schemes (Fleming (2010: 181)).
90 See, e.g., Keith (2000: 250). See Keith (2000: 257–58) for a taxonomy of core appraisals of climate change and deliberate climate modification.
91 Keith (2000: 250).
92 [93] Keith (2000: 250).
93 US National Research Council (1965)
94 Keith (2000: 250).
95 Keith (2000: 252); Fleming (2010: 6); Hulme (2014: 71f.).
96 Miller and Edwards (2001: 143f).
97 See Keith (2000: 251). US government reports monitored Soviet activities and estimated their progress (see, e.g., US National Research Council (1965: 17)).
98 US National Research Council (1965: 10f).
99 US National Research Council (1965: 27).
100 US National Research Council (1965: 27).
101 US National Research Council (1965: xi).

4
YEARS OF FRACTURE

IN 1965, THE PRESIDENT'S SCIENCE ADVISORY COMMITTEE (PSAC) PUBLISHED a report entitled 'Restoring the Quality of Our Environment'.[1] The report ushered in a new chapter in the career of climate engineering. It broke new ground as it prepared a momentous departure from the problematisation of climatic change that we explored in the previous chapter, and that had defined much of the 1930s through 1960s. Instead of problematising human-induced climatic change as a matter of inadvertent and deliberate climate modification, this report introduced the distinction of *problem and response* for the first time. This implied a two-fold shift – a shift that essentially prepared the grounds of the debate over climate engineering today.

What used to be discussed as *inadvertent* climate modification, as the mere side-effect of deliberate modification schemes, now became formulated as an issue in its own right. It became understood as an environmental problem of global societal significance, moving much closer to an understanding of what we discuss today as global warming. And, what used to be discussed as *deliberate* climate modification was no longer merely formulated as a great opportunity to improve the Earth's climate or to mitigate natural hazards. But it appeared as a potential response measure against the very problem of inadvertent modification. It gained significance as a potential remedy against anthropogenic climate change, thus moving much closer to an understanding of what we discuss today as climate engineering.

Specifically, the President's Science Advisory Committee warned that '[t]he climatic changes that may be produced by the increased CO_2 content could be deleterious from the point of view of human beings'. The committee then continues to advise President Johnson to explore measures of deliberate climate modification as a potential response to address this very problem. The authors continue:

The possibilities of deliberately bringing about countervailing climatic changes therefore need to be thoroughly explored. A change in the radiation balance in the opposite direction to that which might result from the increase of atmospheric CO_2 could be produced by raising the albedo, or reflectivity, of the earth. Such a change in albedo could be brought about, for example by spreading very small reflecting particles over large oceanic areas. The particles should be sufficiently buoyant so that they will remain close to the sea surface and they should have a high reflectivity, so that even a partial covering of the surface would be adequate to produce a marked change in the amount of reflected sunlight.[2]

Characteristic for the time, the authors focused primarily on measures to alter the Earth's reflexivity – suggestions that are still currently discussed as solar radiation management (see Chapter 2). What seems noteworthy from our perspective today is that the experts did not refer to the possibility of reducing fossil fuel emissions as a potential response against climate change. Instead, climate engineering appears as the sole suggested remedy here.[3]

This PSAC report was both path-breaking and ahead of its time. In the following years, only part of the anticipated shift fully materialised: anthropogenic climate change did become established as a societal challenge with global political relevance. This did not, however, further fuel political excitement for climate modification efforts, as one might have assumed, given the PSAC's advice. To the contrary, the US government drastically cut research funds to this area during the early 1970s.[4] Climate engineering, in other words, did not emerge as a 'Plan A' in the face of this newly problematised challenge. These measures, in fact, would come into full swing only much later, during the early 2000s (see also Fig. 2.1).

What happened? How can we make sense of this dynamic?

This chapter unpacks how the particular politicisation of anthropogenic climate change during the 1970s through the 1990s re-defined established alliances between climate science and the state, and therein shaped the career of climate engineering for decades to come.

First, we will see how anthropogenic climate change gained political traction during these years as an issue of environmental safeguarding. This meant that climate change became understood as a challenge to reduce rather than expand techno-scientific intervention capacities. It became understood as a challenge that marked the limits of human control over the climate, therefore effectively curbing the earlier political excitement over the prospect of deliberate climate modification that had defined the 1930s through 1960s. As a result, this politicisation of climate change engendered a substantial fracture in established alliances between climate science and politics; it corresponded to a defining shift in the status of climate science for the state. Climate science no longer seemed to promise a potential tool of control at the hands of the state, but, to the contrary, it seemed to question the very hopes of control that had defined the political cultivation of climate expertise until then.

The politicisation of climate change, however, not only fractured established alliances between climate science; it also cemented new ones. In the second part of this chapter, we will see how climate science became established as the *problem-defining authority* for this newly politicised issue and how this new role of climate science as problem-defining authority provided the essential breeding ground in which notions of techno-scientific climate intervention would, eventually, begin to prosper again. The second section of this chapter turns in more detail to this newly emerging role of climate science in the state. We will unpack the defining *expert infrastructure*, as well as the relevant expert *modes of observation* that were essential in assembling this newly politicised issue of climate change (see Introduction for an overview of these concepts). And we will explore how, corresponding to this new problem-defining status of climate science, climate engineering schemes changed their status. While they no longer appeared as an exciting prospect of human control over the atmosphere, these measures did not vanish either. Instead, we can trace how they began to cautiously appear as a potential science-based remedy against this newly defined CO_2 problem.

ENVIRONMENTALISM AND THE POLITICISATION OF CLIMATE CHANGE

We will begin in the following by taking a closer look at the spectacular rise of environmentalism, beginning in the 1960s and 1970s. This rise of environmentalism will help us make sense of why climate engineering lost political traction precisely when climate change gained traction as an issue of global political significance. Specifically, we will see how this rise of environmentalism provided the relevant historical context and defining breeding ground which at once shaped the politicisation of human-induced climate change and questioned earlier hopes of techno-scientific control over the climate.

The rise of environmentalism

In the decades from the 1960s to the 1980s, public awareness and political attention focused increasingly on environmental challenges. This rise of environmentalism materialised very prominently in the social movements of the time. Propelled by the experiences of the Second World War, the Cold War, and the Vietnam War, these movements called for peace and environmental protection. What united such calls for peace and environmental protection was an experience of the grim side of human efforts to control. The experiences of environmental degradation and the atrocious experiences of war – particularly exemplified in the horrid images that reached the world from the Vietnam War – emphasised the vulnerability of societies to the adverse consequences of their own technological prowess. In the fall of 1969, the New York Times predicted the force of this set of newly emerging issues on the horizon:

> Call it conservation, the environment, ecological balance, or what you will, it is a cause more permanent, more far-reaching, than any issue of the era- Vietnam and Black Power included.[5]

This rise of environmental concerns eventually began to seep into the political sphere as states and international consortia began adopting the issue and internalised it in governance structures and legislative orders. 'Green' parties and political programs began proliferating, and between 1972 and 1982, the governments of 118 countries established agencies to deal with environmental issues.[6]

Environmental legislation enjoyed strong bipartisan support. This is particularly noteworthy for the case of the United States where the outright rejection of the very existence of the climate change issue would later emerge as a legitimate political position. In August 1969, a Senator from Alaska complained that 'suddenly out of the woodwork come thousands of people talking about ecology'.[7] Towards the end of the 1960s, President Nixon was confronted with an 'extraordinary outburst of mass public pressure'[8] and 'massive bipartisan popular demand'[9] to address pressing environmental problems. For the newly elected president, environmental protection provided 'a welcome opportunity'.[10]

It seems difficult to imagine, judging from the deep partisan trenches that define environmental policy in the United States today, that in the 1960s and 1970s, environmental protection emerged as a unanimous consensus issue at a time of deep ideological divisions. In their analysis of conservative environmental policy in the United States, James Morton Turner and Andrew Isenberg emphasise just how differently this political setting was compared to today's situation. In fact, the authors describe the development that would unfold from Nixon to Trump as the 'Republican Reversal', suggesting how conservatives used to play a critical role in driving environmental policy.[11] And so, although Nixon had virtually no prior record on environmental issues and had not run on them for office, he grasped this opportunity. In his 1970 State of the Union, Nixon expressed his desire to turn the 1970s into the historical period, where '[we] transform our land into what we want it to become'.[12] In April of that year, the first Earth Day amassed more than 20 million US Americans who demonstrated for environmental protection. A couple of weeks later, Nixon formally signed the National Environmental Policy Act (1970) into law, which created the Environmental Protection Agency (EPA). In his retrospective on *The EPA at 40*, Richard Andrews argued that '[i]t is fair to say that President Nixon saw a mob coming, jumped in front of it and called it a parade'.[13] The Wilderness Act (1964), the National Environmental Policy Act (1970), the Clean Air (1970) and Water (1972) Acts, and the Surface Mining Control Reclamation Act (1977) were but a few examples of the many laws passed with strong bipartisan support in both the House and the Senate during these years.[14] During this time, environmentalism resonated with both conservatives and liberals as it touched on 'anxieties about pollution and overpopulation' and 'desires for a clean and aesthetically pleasing environment'.[15] These bills implemented environmental 'legal trumps', laws that prioritised the protection of the environment over other potentially conflicting interests (e.g., commercial interests).[16] These bills thus further 'stacked the deck' of the environmental movement and importantly institutionalised environmental protection as an issue within the political system.[17]

The politicisation of climate change

This spectacular rise of environmentalism during the 1960s and 1970s provided the essential breeding ground that defined the politicisation of climate change that began in the 1980s. This, in other words, was the particular historical setting that climate change was born into, when Jim Hansen, a scientist at the National Aeronautics and Space Administration (NASA) testified before Congress that 'the earth is warmer in 1988 than at any time in the history of instrumental measurements', and that this warming was now large enough to be ascribed to the so-called greenhouse effect.[18] That same day, the New York Times reported on the incident, announcing that '*Global Warming Has Begun*'.[19]

The media and outspoken scientists added to a 'public imagination' of the 'global climate catastrophe',[20] effectively pushing the issue into the political sphere, not only in the United States. While scientists had problematised individual dimensions of human-induced climate change over the preceding decades already (see Chapter 3), it was not until the late 1970s (and especially the 1980s) that human-induced climate change was increasingly being problematised as an environmental issue of global political and societal significance.[21] The notion of inadvertent climate modification which had defined the climate change debate during the 1960s was thus incrementally abandoned and replaced by the notion of 'global warming' – 'a long-expected global warming trend linked to pollution' as the New York Times put it.[22] In their account of The Globalization of Climate Science and Climate Politics, Clark Miller and Paul Edwards suggest how, in contrast to earlier decades, climate change was no longer primarily conceptualised in its plural form – as climatic changes concerning the meteorological conditions of individual places and geographic regions. Instead, scientists increasingly began to devise a global category, 'something more closely akin to the global environment: a natural object to be understood, investigated, and managed on planetary scales'.[23]

The years following the summer of 1988 saw a growing and consistent public sense of urgency regarding this issue. Ann Keller recounts that, in 1987, Congress held only four hearings on climate change in four separate committees. One year later, it was already present in nine hearings in eight separate committees, and in 1989, a total of 21 hearings were held in twelve committees.[24] In addition, countless organisations were established, programs coordinated, and reports written up in the name of tackling this challenge. Climate change became part of a broader agenda focused on efforts such as increasing environmental protection, stopping rainforest depletion, restoring biodiversity or, more generally, achieving an ecological balance between nature and society.[25]

…And the fate of climate engineering: Fractured alliances between climate science and the state

This problematisation of climate change engendered a substantial shift – a 'fracture' – not only in the career of climate engineering, but also in established alliances between climate science and the state more generally. To begin with, the spectacular rise of environmentalism during the 1960s and 1970s effectively marked the end of techno-optimism, which had shaped the war and after-war years. Instead, awareness of technological risks to society rose, as we have seen at the outset of this section. Incidents such as US cloud seeding activities during the Vietnam war, and later the nuclear accident in Chernobyl in 1986, emphasised the vulnerability of societies to the adverse consequences of their own technological progress.[26] As a result, earlier excitement about visions of technological climate and weather control lost traction.[27] Esteemed climate scientists, such as Stephen Schneider and William Welch Kellogg warned that the sudden scientific progress in climate science would not ultimately result in options of control. In a 1974 *Science* article on 'Climate Stabilization',[28] the authors challenge the assumption that (climate) prediction enables (climate) control: 'even if we could predict the future of our climate, climate control would be a hazardous venture'.[29]

This particular politicisation of climate change as an environmental challenge impinged on existing alliances between climate science and the state. The massive climate science infrastructure, which had initially been developed as a tool for the state, and established to address geopolitical and military objectives, now 'discovered a possibility' (namely anthropogenic climate change).[30] Born into the historical context of environmentalism, this 'possibility' gained political steam and societal attention as a challenge that threatened to undermine the global political and economic status quo.[31] Zeke Baker described this as a 'historical inversion' of the situation established during and after the years of World War I and II.[32] While the geopolitical challenges of the 1930s through the 1960s had aligned meteorological and political agendas, the environmental challenges of the 1960s through the 1980s provoked a division between scientists on the one hand, and political and economic elites on the other. Prominent figures such as Rachel Carson, Paul and Anne Ehrlich, Edward Wilson, Paul Raven and Jim Hansen represented a 'new breed' of scientists who spoke to a broader audience and 'raised the alarm' about environmental challenges.[33]

CLIMATE SCIENCE AS A PROBLEM-DEFINING AUTHORITY AND THE RE-NORMALISATION OF CLIMATE ENGINEERING

The previous section of this chapter embedded the politicisation of the climate change issue as we know it today in its historical setting. We have seen how climate change – now in its singular form, understood as global warming – gained political traction in the 1980s, following a spectacular rise of environmentalism during the preceding decades. This context explained part of the puzzle, raised at the outset of this chapter. Specifically, this politicisation of the climate change issue helped explain why climate engineering lost political traction in the 1970s and why this response initially did not fit the problem (just yet). We saw how this politicisation of climate change actually fractured established alliances between climate science and the state.

In the second part of this chapter, we will complement this picture and further unpack the status of climate engineering in the early US political exploration of the climate change issue. As much as the politicisation of climate change provided a fracture and eroded established alliances between climate science and politics, it gave rise to powerful new ones. In what follows, we will see how climate science became established as the *problem-defining* authority for the politicisation of climate change. We will see how this newly emerging role of climate science in the state provides the precise starting point of a kind of re-normalisation of climate engineering in US climate policy. It explains the context in which these measures incrementally began to prosper again.

Cautious beginnings: Climate engineering (re-)enters US climate change policy

Let's begin by taking a closer look at the document corpus, comprising all US policy records, referring to or referencing climate engineering from the Federal Digital System (see Introduction and Appendix for tabular overview of the corpus). As we have seen in Chapter 1 already, this database suggests that climate engineering played virtually no role in early US climate change policy (see Fig. 2.1). Climate engineering entered the political exploration of this newly politicised issue only in May 1990. An isolated reference to Cesare Marchetti's seminal 1977 paper – '*On Geoengineering and the CO_2 Problem*',[34] that had coined the term 'geoengineering' – appeared on the US political record for the first time, hidden in a 508-page-long report by the Intergovernmental Panel on Climate Change (IPCC), and tucked away in the archives of the National Oceanic and Atmospheric Association (NOAA).[35] During the years that followed, notions of climate engineering remained essentially invisible in the political exploration of climate change in the United States, confined to the outer margins of the examined policy documents.

These observations essentially speak to what we saw in the previous section. Although climate change had already emerged as a relevant policy concern on the US political agenda some decades earlier, there was hardly any reference to climate engineering in this context until the turn of the new millennium (see Appendix). During the 1990s, there are only two policy documents that touched on climate engineering as a potential response measure to climate change at all, and they did so only indirectly.[36]

So, how *did* climate engineering matter here? By taking a closer look at these few references to climate engineering from the 1990s, we might begin to understand how these measures made their way back onto policy agendas in the years that would follow.

What stands out in this context is that climate engineering entered the US political debate over global warming as the literal footnote to a controversial scientific hypothesis. In other words, it emerged in reference to a scientific puzzle. These policy documents suggest how a 'discernible human influence' on the climate had sparked a policy debate which questioned the very grounds of the scientific method and marked its epistemological limits.[37] Chairman John Chafee opened his 1997 hearing on 'Global Climate Change' with a swift historical excursion on the scientific 'discovery' of anthropogenic climate change:[38] '*what's going on here? What are the scientists saying*'? The hearing was the first in the United States to contain a reference to 'geoengineering' (see Appendix). Somewhat ironically, the reference came from Stephen Schneider, professor at Stanford University and one of the most influential climate scientists of the time, who we met in the previous section of this chapter as one of the critical voices on climate engineering. Schneider had warned already in the 1970s that visions of climate control would be 'a hazardous venture', as we have seen.[39] In this 1997 hearing on 'Global Climate Change', Schneider now added a list of climate policy recommendations by the National Academies to his testimony, which suggested research into climate engineering as one among various approaches to address the issue of climate change.[40] This list of recommendations had been prepared some years earlier, in 1992, in the context of an assessment of the 'Policy Implications of Greenhouse Warming'.[41] The section of the report, which was quoted in the 1997 hearing suggests to 'undertake research and development projects to improve our understanding of both the potential of geoengineering options to offset global warming and their possible side-effects'. The quoted section continues to stress that 'this is not a recommendation that geoengineering options be undertaken at this time, but rather that we learn more about their likely advantages and disadvantages'.[42] Climate engineering emerges here as an approach that should be researched, but not implemented, much akin to the dichotomies that define the debate today (see Introduction).

The context of these sparse and marginal references to climate engineering complements the picture that the first section of this chapter painted. It suggests just what a difference the politicisation of climate change had made to the US political exploration of climate intervention measures. Instead of fuelling hopes for techno-scientific control, climate engineering appeared here in the midst of a policy struggle over scientific facts regarding anthropogenic climate change. It emerged in debates that concerned the scientific grounds and epistemological premises of climate change. This context thus hints at the emerging new role of climate science in the state that came after the 'fracture'. While climate science no longer seemed to primarily promise a tool of control, it became established as the *problem-defining* authority for the politicisation of climate change. Climate science, in other words, became essential in assembling climate change as a governance object. It played a critical role, not only in regard to placing climate change on the political agenda, but primarily in defining the terms in which the issue was problematised and addressed. It structured the terms in which this newly raised issue became legible as a political issue in its own right.

In the rest of this section, we will contextualise this initial picture, drawn from the sparse references to climate engineering in the document corpus prior to 2000. By turning to existing scholarship and politically commissioned scientific assessment reports from the time, we will see how this new role of climate science as a problem-defining authority provided the essential breeding ground in which notions of deliberate climate intervention would slowly begin to prosper again. It is this very context in which the current debate over climate engineering has its direct roots. Climate engineering makes its debut here as a potential response to a newly assembled problem. And in this particular form, it sets out on a journey of re-normalisation.

Forging new alliances between climate science and the state

The politicisation of climate change cemented new alliances between climate science and the state along at least two critical channels. On the one hand, climate science shaped the emerging politics of climate change via individual experts and rather informal networks of scientific experts. On the other hand, the political system began institutionalising climate expertise within the state: politics, in other words, began internalising climatological modes of observation in making sense of and addressing this newly raised issue. As we will see, these two channels are not only similar, but in a sense, they provide the very basis of the arenas in which science came to shape the emerging politics of climate engineering since around 2009, as the book has already explored (see Chapter 2).

To begin with, climate scientists appeared as important spokespeople on the urgency of climate change. During the 1970s, and continuing up to the 1990s, individual experts, such as James Hansen or Stephen Schneider, for example, reached a broad audience via the mass media, and therefore, indirectly, they also reached politics.[43] With this visibility, these experts played a critical role in communicating the issue of climate change and pushing it into the sphere of politics. Climate scientists began joining leaders of environmental movements in a coalition that effectively placed climate change as a global environmental problem on political agendas and would eventually help institutionalise the issue within the federal bureaucracy.[44] In other words, climate change became internalised by the political system as the result of a scientific 'push' rather than a 'pull' from elected decision-makers.[45] The literature suggests that this push came from a well-defined group of scientists with overlapping institutional affiliations; a group that was described as a 'nonsinister conspiracy' advancing their climatological agenda.[46]

Notably, this metaphor of a kind of scientific 'conspiracy' is a recurring theme that describes the pronounced role of individual scientific experts or informal networks of scientists, pushing their perspectives on emerging political agendas. In accounts of the politicisation of biodiversity loss, for example, we find the notion of a 'mafia': researchers such as Rachel Carson, Paul and Anne Ehrlich, Edward Wilson, or Paul Raven were not only scientifically respected, but also publicly outspoken and institutionally well connected.[47] Wilson in this context recounts belonging to a group of biologists that he jokingly referred to as 'the rainforest mafia', suggesting the relevance of these scientists in advancing their perspectives in the political realm.[48] And in the case of climate engineering, the notion of a 'clique' has been advanced to capture the powerful role of a distinct group of scientific experts, advancing and shaping the programmatic exploration of climate engineering as an issue in its own right (we will come back to this in Chapter 6). While the scientists of course vary in each case, these notions of non-sinister 'conspiracies', 'mafias', or 'cliques' of scientists point to the critical role of highly visible scientific experts in channelling scientific observations into the political realm. In the case of climate change, this 'non-sinister conspiracy' was essential in making the issue legible to politics; it assumed a critical role in assembling this issue and giving it concrete shape.

Once the issue of climate change had seeped into the political realm – reinforced by a push coming from scientific spokespersons – politics began taking matters into its own hands and internalised the issue within the federal bureaucracy. We might think of this process as a kind of political adoption or translation of the issue of climate change. As early as 1977, the National Academies called for the need to restructure the science-politics nexus at the national level across established infrastructures, such as disciplinary boundaries or bureaucratic programs to properly tackle climate change. The National Academies suggested the need to 'weave together the interests and capabilities of the scientific community and the various agencies of the federal government in dealing with climate-related problems'. It argued that properly tackling these problems 'will involve coordination of research in many scientific disciplines and … require adjustments in national policy or the formulation of new legislation'.[49] This political internalisation of climate change thus rested essentially on the search and establishment of problem-relevant expertise.

In the United States, the history of the United States Global Change Research Program (USGCRP) demonstrates particularly clearly how climate science became institutionalised as the problem-defining authority on global climate change. As we will see shortly, this history brings us back to the set of acronyms that would later define the inventory of climate engineering expert capacities within the state (Chapter 2). This institutionalisation of climate science as the problem-defining authority provided a relevant foundation for the recent renaissance of climate engineering. It established a problematisation of climate change that climate engineering promised to respond to.

In the following, we can draw on Roger Pielke Jr.'s highly instructive two-part history on the establishment of the US Global Change Research Program to reconstruct how this program came into existence. As extreme weather events hit different world regions throughout the 1970s, US Congress was confronted with calls to increase climate science funding.[50] Congress saw the need to federally fund climate science as a means to improve weather predictions and alleviate the consequences of such incidents.[51] Eventually, this congressional initiative led to the National Climate Program, a direct predecessor to the US Global Change Research Program. The National Climate Program was signed into law during the fall of 1978 as an inter-agency program to be coordinated within the National Oceanic and Atmospheric Administration (NOAA). Its main purpose was to conduct climate research to assess the policy relevance of this newly emerging issue.[52] As a result, the participating agencies – particularly the National Aeronautics and Space Administration (NASA), NOAA, and the National Science Foundation (NSF) – began fostering climate change capacities and developing specific areas of climate expertise, respectively. For these agencies, a joint global change agenda promised organisational stability by warding off budgetary cuts or political assaults.[53] NOAA's critical role as a 'home for climate change research' within the state was further consolidated when the Reagan administration tasked NOAA administrator Anthony Calio in 1986 with heading a White House Domestic Policy Council working group on climate change.[54] The working group's purpose was to assess the problem, advance an integrated agenda on the issue within the state, and suggest what the President was 'supposed to do about it'.[55] Finally, the Office of Management and Budget (OMB) also pushed for a coordination of federal research efforts on climate change. As a result, science advisor William Graham decided to form the Committee on Earth Sciences in 1987. After some hiccups, this committee prepared the establishment of the US Global Change Research Program two years later in form of a budget summary. In essence, the USGCRP 'began as a multi-agency budget [...] crosscut', a funding table organised by agency and discipline (or program).[56] The program was then effectively established with the passage of the Global Change Research Act of 1990, which changed the mandate of the former Committee on Earth Sciences to go beyond the mere coordination of research budgets.[57] The purpose of the USGCRP was to provide for development and coordination of a comprehensive and integrated United States Research program which will assist the Nation and the world to understand, assess, predict, and respond to human induced and natural processes of global change.[58]

The US Global Change Research Program thus sought to translate the issue of climate change into a manageable political challenge by means of climatological expertise. It institutionalised climate science to not only 'understand' and 'assess', but also 'predict and respond to human induced and natural processes of global change'.[59] The program sought to match scientific efforts of understanding climate change directly to political efforts aimed at governing climate change. It set out to 'develop a predictive understanding of the earth's climate', with policymakers expecting it to deliver 'action programs that are rational and sensible and cost effective'.[60]

These expectations were, of course, not met. Nonetheless, with the benefit of hindsight, we can see that this program established climate change as a political issue that would coordinate research agendas across federal agencies for decades to come.[61] Writing almost twenty years ago, Roger Pielke Jr. asserted that the legacy of the National Climate Program's efforts in the late 1970s to define agency roles within global climate change[62] extended into the 1990s, via the US Global Change Research Program. Now we can see that, in fact, its legacy extended well into the teens of the new millennium. Through the lens of these expert agencies, climate engineering would emerge as yet another chapter in the federal institutionalisation of climate change expertise in the US, which, at its core, has evolved around the same agencies since the 1970s.

At the international level, it was the establishment of the Intergovernmental Panel on Climate Change (IPCC) which cemented the political problem-defining role of climate science most prominently. The IPCC was set up by the World Meteorological Organization (WMO) and the United Nations Environmental Programme (UNEP) in 1988, a year that marked the height of climate change politicisation. Ever since then, the IPCC emerged as one of the most prominent and controversial organisations, specifically initiated to bundle policy-relevant expertise on anthropogenic climate change. Its mission has been 'to provide the world with a clear scientific view on the current state of knowledge in climate change and its potential environmental and socio-economic impacts'.[63] Set up as a representative parliamentary body, the IPCC essentially embodied the newly emerging alliances between climate science and climate politics. The organisation was established and gained global status during a time when the reality and severity of global climate change was primarily addressed as a scientific challenge. In the following decades, it 'gradually acquired status and authority' in structuring global problem observations of climate change, and more specifically, in linking scientific and political observations on the challenge of climate change.[64] The history and dynamic global status of the IPCC thus reflect the inter-governmental institutionalisation of policy-relevant expertise on climate change, its specific scope and outlook more broadly.

Summing up, the 1970s through 1990s demonstrated how the politicisation of climate change not only disrupted existing alliances between climate science and politics, but also how it effectively sealed new ones. These years introduced two channels in particular that established the new – problem-defining – role of climate science in US politics. On the one hand, the prominent role of individual scientists, such as James Hansen or Stephen Schneider, suggested the power of informal networks of scientific experts in shaping political agendas. The establishment of the IPCC and the formation of the USGCRP, on the other hand, demonstrated the critical role that the targeted political organisation of scientific expertise played in this context. Climate change effectively established new boundary organisations (e.g., the IPCC), and it differentiated existing structures, coordinating, for example, a unified policy agenda across a myriad of diverse existing agencies in the United States, particularly within the departments of agriculture, energy, and state, the Environmental Protection Agency (EPA), NASA, NOAA), the NSF, and the US Geological Survey (USGS). As we will see in the following chapters, both mechanisms have continued to shape the career of climate engineering in the United States for years to come. They constitute an essential component of the climate engineering expert infrastructure (see particularly Chapter 6).

Assembling climatic change: From inadvertent and deliberate modification to managing the greenhouse problem

In the final two sections of this chapter, we shift our gaze from the structural dimension of the science-state alliances that defined this historical setting of the career of climate engineering to the epistemic dimensions of these alliances. Connecting to the analytical framework I outlined in the Introduction, this means that we are changing our perspective from scientific experts to scientific expertise – from examining the *expert infrastructure* to examining the *expert modes of observation* that shaped this 'stage' in the career of climate engineering and undergirded this new problem-defining authority of climate science in the state.

During the late 1970s, at a time when the Carter administration sought to utilise coal for tackling the oil crisis, it was carbon dioxide that moved into the centre of climate policy attention.[65] While scientists had explored the role of CO_2 on the climate for some decades, it was only now that this issue fully arrived in the political arena.[66] Policymakers began commissioning assessments that increasingly focused on atmospheric chemistry and carbon dioxide pollution as a source of 'Greenhouse Warming'.[67] In these final two sections of the chapter, we will see how, during the 1970s through 1990s, climate engineering – although not pursued as 'Plan A' – incrementally began to emerge as a potential response to this newly assembled 'carbon dioxide problem'.[68] Specifically, we will see how these measures catered to different visions of tackling this newly assembled challenge, and in extension, to different visions of the role of climate science in the state.

We will begin this section by unpacking how notions of techno-scientific climate intervention were couched in the very formulation of the 'Greenhouse Problem'. By turning to three politically commissioned scientific assessment reports by the US National Academies, we will see how these measures appeared as a potential answer to the issue of managing atmospheric chemistry, and specifically, how they catered to environmental concerns in this context. This 'carbon dioxide problem' no longer appeared as the mere downside of deliberate climate modification schemes, as I suggested earlier, but it was cast as an issue of environmental safeguarding. Concerns over what an 'optimum global climate'[69] or a 'good environment'[70] that humans should aspire to might look like began to guide expert observations.

In a 1977 expert assessment on 'Energy and the Climate', the National Academies, for example, suggested that policymakers and scientists should address the question of '[w]hat [...] the atmospheric carbon dioxide content [*should*] be over the next century or two to achieve an optimum global climate'.[71] The report discussed measures to remove carbon dioxide from the atmosphere as well as measures to reflect sunlight back to space as a potential means for achieving this goal.[72] Climate engineering thus emerges here as part of an agenda that envisions climate science as catering to environmental concerns over this newly politicised issue of global warming. Notably, this implies that these measures no longer appear as a straightforward techno-fix. So, while the authors acknowledge that counteracting 'the climatic effects of added carbon dioxide in the air' might be achieved by 'increas[ing] the albedo, or reflectivity, of the earth and thereby [...] reduc[ing] the incoming solar radiation', they also conclude that 'no practical, plausible, and reliable means to accomplish such an increase seem to be at hand'.[73] Climate engineering emerges here as part of a kind of climate science that not only defines and monitors this environmental issue of global warming, but potentially also manages it.

Building on these kinds of observations, President Carter signed the Energy Security Act in June 1980, calling for a 'comprehensive assessment of the implications of increasing carbon dioxide due to fossil fuel use and other human activities' by the National Academies and the Office of Science and Technology Policy (OSTP).[74] In response to this, the National Academies formed a Carbon Dioxide Assessment Committee under its Climate Board in 1980. In 1983, this committee published what can be understood as one of the earliest policy frameworks for tackling anthropogenic climate, and climate engineering was part of it. Specifically, this 'framework for policy choices' presented a four-tiered taxonomy of policy options to address 'CO_2-induced climatic change'. Between the two poles of 'reduc[ing] CO_2 production (1)' or 'adapt[ing] to increasing CO_2 and changing climate (4)', the framework listed two more response categories, namely the option to 'remove CO_2 from effluents or atmosphere (2)' and to 'make countervailing modifications in climate, weather, hydrology (3)'.[75] This taxonomy thus provides an early version of how climate policy approaches largely continue to be categorised and discussed today, namely by distinguishing between mitigating, adapting to, or intervening in climate change.

Finally, around a decade later, the National Academies presented another report addressing the 'Policy Implications of Greenhouse Warming'. We encountered this report earlier during this chapter when it became part of the very first reference to 'geoengineering' in the context of a congressional hearing in the United States. It was Stephen Schneider who added some of the text's climate policy recommendations to his congressional testimony in 1997. This report listed climate engineering measures as a potential policy approach 'to combat or counteract the effects of changes in atmospheric chemistry', and in fact, featured an entire chapter on these measures.[76]

These three reports not only document the political relevance of climatological expertise in assembling climate change as a governance object, but they also demonstrate how this epistemology of 'the carbon dioxide problem' immediately implied climate intervention measures as a potential remedy. These texts and taxonomies can be understood as providing the immediate roots of today's US political exploration of climate engineering. First, they suggest how climate engineering slowly and cautiously became understood as a potential response to anthropogenic climate change. While the tone of these expert assessments is notably less optimistic than during the 1960s, climate engineering becomes part of an agenda here that envisions climate science as catering to environmental concerns in defining, monitoring, and potentially also managing the climate change issue. This new understanding of climate engineering is also reflected in scientific publications of the time. In 1996, the publication Climatic Change, for example, featured an issue that asked if we 'may', 'could' and 'should engineer the Earth's climate?', focusing notably on normative questions.[77]

And secondly, these texts and taxonomies provide a critical episode in the 'career' of climate engineering because they integrate what we discuss today as carbon dioxide removal and solar radiation management into one policy framework. These documents make measures to remove carbon from the atmosphere and measures to alter the Earth's albedo into distinct components of one (climate) policy agenda. In their promise to tackle this issue, these two sets of otherwise fundamentally different techno-scientific concepts are united, which is an essential step towards the emergence of climate engineering as a political issue in its own right.

We may contextualise this latter point with a brief excursion to the academic debate surrounding climate engineering at the time. This differentiation of carbon dioxide removal and solar radiation management as two distinct measures tackling one societal issue was prepared in two path breaking scientific contributions from 1977. These contributions were path breaking in the literal sense of being commonly perceived today as the direct scientific origins of the current climate engineering debate. Each of these texts introduced one of the measures respectively, without, however, relating them yet. In 1977, Russian geoscientist Mikhail Ivanovitch Budyko provided one of the first suggestions of what is discussed today as solar radiation management.[78] In his book, *Climatic Changes*, Budyko suggested applying a stratospheric layer of aerosols, reflecting incoming sunlight back to space to stabilise current climatic conditions.[79] The suggestion of what is discussed today as carbon dioxide removal came from Italy during that same year. Italian physicist Cesare Marchetti coined the term 'geoengineering' in the early 1970s and then formally introduced the concept in the inaugural issue of *Climatic Change* in 1977.[80] We stumbled across this paper earlier as it provided the very first reference to 'geoengineering' in the corpus of climate engineering-relevant policy documents, hidden in a footnote of a 508-page long report from 1990.[81] In this contribution, Marchetti suggests 'geoengineering' as a means of '[tackling] the problem of CO_2 control in the atmosphere'. Climate change emerges here as a (CO_2) pollution issue and Marchetti proposes to address this issue with 'a kind of "fuel cycle" for fossil fuels where CO_2 is partially or totally collected at certain transformation points and properly disposed of'. For this 'disposal' of CO_2, Marchetti primarily explores ocean-based methods. Specifically, the paper suggests that

> CO_2 is disposed of by injection into suitable sinking thermohaline currents that carry and spread it into the deep ocean that has a very large equilibrium capacity. The Mediterranean undercurrent entering the Atlantic at Gibraltar

has been identified as one such current; it would have sufficient capacity to deal with all CO_2 produced in Europe even in the year 2100.[82]

Ocean-based measures for removing CO_2 from the atmosphere received increasing attention during the decades following Marchetti's paper. Presumably fuelled by the increasing political interest in 'managing' the CO_2 problem, the 1990s witnessed a wave of ocean fertilisation studies, effectively ushering in the beginning of intense international studies of climate intervention measures.[83]

Summing up, these politically commissioned assessments and scientific analyses suggest how hopes of techno-scientific management and control continued to structure the newly forged alliances between climate science and the state. Their point of reference, however, had substantially changed. What was seen during the 1930s through 1960s as a great possibility for national strategic and military application, now cautiously (re-)appeared as a potential remedy against a problem of environmental safeguarding. In these accounts, measures of techno-scientific climate intervention emerged as an option to manage atmospheric chemistry and establish an 'optimal climate' to counteract CO_2 pollution, and to tackle greenhouse warming.

Economic modes of observation: Assembling climatic change as an issue of 'just spending money'

In the final section of this chapter, we will see how this climatological gaze onto the 'carbon dioxide problem' and its potential 'management' via climate engineering measures spoke not only to environmental concerns of safeguarding nature. Beginning in the 1980s and 1990s, it was also economic concerns that prominently structured politically commissioned expert assessments on this topic. These economic modes of observation mobilised measures of climate engineering in an effort to speak to a rather different constituency than the environmental movement. These observations suggested climate engineering as a means of decoupling efforts to address climate change from interventions in the economic and political status quo. Climate engineering thus becomes part of a very different vision for climate science in the state here. Instead of raising the alarm about issues of pollution and ecological safeguarding and marking the limits of growth and techno-scientific control, these economic analyses paint a picture of climate science as a tool that permits tackling this issue while stabilising the economic and political status quo.

Again, we can turn to a number of politically commissioned assessment reports by the National Academies to illustrate this point, some of which we touched upon in the previous section already. Thomas C. Schelling, Nobel laureate and pioneer in game theory, for example, provided the final chapter of the aforementioned 1983 National Academies' report on *A Changing Climate*. Schelling's chapter explored the 'Implications for Welfare and Policy' that the issue of climate change raised.[84] The text builds on climatological modes of assembling this greenhouse problem to then cast the issue of climate change in economic terms. Schelling argues that '[w]arming the atmosphere currently is more economical than cooling it, because it happens as a byproduct of energy consumption that would be costly to reduce or terminate'.[85] This kind of problematisation deviates from the common environmental problematisations of climate change. The key question here is no longer what the carbon dioxide content of the atmosphere should be or what an 'optimum climate' would look like, but rather what these things would cost. Schelling then mobilises the notion of climatological intervention not as a science-based means for halting pollution and managing an 'optimum' atmospheric carbon composition, but as an economical way of addressing this issue:

> But we know that in principle cooling could be arranged. Volcanic eruptions have done it. [...] [W]e should not rule out that technologies for global cooling, perhaps by injecting the right particulates into the stratosphere, perhaps by subtler means, will become economical during coming decades.[86]

A decade later, the 1992 National Academies' report on 'Policy Implications of Greenhouse Warming' again emphasised 'the relatively low costs at which some of the geoengineering options might be implemented'[87] – an insight that is readily quoted in recent explorations of the topic.

A couple years before being invited to co-author the 1992 National Academies report, William Nordhaus criticised too much action against climate change[88] and stressed that 'research on climate engineering may well be the best investment'.[89] And Thomas Schelling presented his 'Economic Diplomacy of Geoengineering'[90] in 1996, praising how climate control would able to 'immensely simplify greenhouse policy, transforming it from an exceedingly complicated regulatory regime to a problem in international cost sharing [...]'.[91] Again, climate engineering is mobilised here in an effort to transform the political economy of the climate change issue:

> Putting things in the stratosphere or in orbit can probably be done by 'exo-national' programs, not depending on the behavior of populations, not requiring national regulations or incentives, not dependent on universal participation. It will involve merely deciding what to do, how much to do, and who is to pay for it.[92]

From this perspective, tackling climate change would hardly require questioning the status quo, but rather would 'involve[s] just spending money'.[93]

To make sense of the status and saliency of such economic expert observations at the time, we shall return to the 'fracture' that the politicisation of climate change had imposed on established alliances between climate science and the state. As I described at the start of this chapter, this fracture resulted from the new problem-defining authority of climate science. With the politicisation of climate change, climate science no longer seemed to provide expertise that promised a tool of political control. To the contrary, by raising the alarm about this newly politicised issue, climate science seemed to directly question the hopes of control and the general techno-optimism that had defined the 1930s through 1960s.

One result of the politicisation of climate change issue in the United States, therefore, was a general decline in bipartisan support for environmental policy. While both the Democratic and Republican parties could take great pride in US environmental leadership through the late 1980s, James Turner and Andrew Isenberg suggest that the politicisation of anthropogenic global warming changed just that.[94] The authors argue that it was Ronald Reagan who initially 'broke' the bipartisan consensus on the need for environmental protection.[95] Ever since then, congressional voting scores on environmental issues have displayed a continually widening gap.[96] The 1992 'Earth Summit' sealed this change of direction for years to come. While the Earth Summit brought forth the United Nations Framework Convention on Climate Change (UNFCCC), the Bush administration refrained from committing to clear emission reduction targets.[97] By the 1990s and early 2000s, US political support for climate change action was dwindling across the political spectrum. Due to opposition from the Senate, Bill Clinton failed to sign the Kyoto Protocol, and by 2001, George W. Bush made it clear that he would neither abide by the protocol, nor had any intention of implementing binding emissions reductions for the United States.[98] Despite rising urgency in the public perception of climate change, US climate policy thus grew increasingly divided, even coming to an effective halt at the federal level.

This brief digression is to emphasise just how consequential the problematisation of climate change as an environmental issue has been in the United States and what lasting effects it had on the political economy of this issue. Against this backdrop, the early economic appraisals of climate engineering can be seen as an effort to question this outlook. In a sense, these economic accounts can be read as an effort to mend the proposed fracture and reinstate climate science as a tool for the state. By presenting climate engineering as an economically smart solution to the 'carbon dioxide problem', the authors suggested a perspective on the climate change issue that sought to reconcile political concerns over global warming with interests in maintaining the economic and political status quo. These economic perspectives are therein essential for understanding the controversial debate over climate engineering today.

On the one hand, these accounts explicate a dimension of the political economy of climate engineering which has been decisive in fostering political support of these measures ever since. They speak to political interests which have been committed to warding off any diagnoses pointing to the need for substantial structural changes to the way modern economies operate, depending on ever-increasing sources of energy. On the other hand, these economic accounts explicate the grounds for the outright rejection of climate engineering measures by many advocates for policy action on climate change today.

When climate intervention measures are advocated as both a technologically straightforward and cheap fix to the climate change problem, they clash spectacularly with calls for more environmental safeguarding, techno-scientific humility, and the limits of human control. These economic accounts thus blatantly explicate what many who are critical of the very idea of climate engineering fear: they mobilise climate engineering as a grounds for divesting policy action on climate change and avert structural change to the existing political-economic system. The kind of narrative that is presented in these economic accounts is therefore essential for understanding why suggestions to deliberately intervene in the climate are so often perceived as radically differing from established perspectives on the climate change issue. This assumption, however, masks the fact that suggestions to deliberately modify the climate have been an integral part of the entire history of what we discuss today as anthropogenic climate change.

NOTES

1. President's Science Advisory Committee (1965).
2. President's Science Advisory Committee (1965: 127).
3. See also Keith (2000: 254).
4. Keith (2000: 253); Kwa in Miller and Edwards (2001: 152).
5. Bendiner qtd. in Lewis (1985: 6).
6. Haas (1992: 9).
7. Lewis (1985: 6).
8. Andrews (2010: 226).
9. Andrews (2010: 223).
10. Andrews (2010: 227).
11. Turner and Isenberg (2018).
12. Lewis (1985: 7).
13. Andrews (2010: 227).
14. See, e.g., Table 1–1 in Turner and Isenberg (2018: 33).
15. Turner and Isenberg (2018: 32).
16. Turner and Isenberg (2018: 21).
17. Weingast (2005: 326).

18 Committee on Energy and Natural Resources, United States Senate (1988: 39). In hindsight, this incident has been stylised as a crucial moment to the career of global warming (see, e.g., Pielke (2010: 9); Fleming (2010: 225); Morton (2016: 91f.)).
19 Shabecoff (1988).
20 Weart (2008: 140). For the rise of climate 'catastrophism' (*Katastrophismus*) in Germany, see, e.g., Weingart, Engels, and Pansegrau (2008).
21 See, particularly, Miller and Edwards (2001); Miller (2004). In his analysis, *Climate Science and the Making of a Global Political Order*, Clark Miller suggests how climate change was but one of several prominent issues that became re-conceptualised during the 1980s and 1990s 'in explicitly global terms' (Miller (2004: 49)).
22 Shabecoff (1988). For a concise overview of this shifting problematisation of climate change, see Keith (2000: 257ff.). For relevant expert assessments from the 1960s, see, e.g., US National Research Council (1965); see also US President's Science Advisory Committee (1965). For relevant expert assessments from the 1970s, see, e.g., US National Academy of Sciences (1977); see also Study of Critical Environmental Problems (1970).
23 Miller and Edwards (2001: 7).
24 Keller (2009: 219).
25 See, particularly, Meadows and Club of Rome (1972). For a critical perspective, see, e.g., Taylor and Buttel (1992).
26 Ulrich Beck published his diagnosis of the risk society during that same year (Beck (1986)).
27 Keith (2000: 253). In his analysis, *The Rise and Fall of Weather Modification*, Chungling Kwa suggested that 'it is tempting to think that at some point [the] practical and theoretical impossibility [of climate modification] imposed itself so strongly on the meteorological community that the field was simply abandoned. This appears not to be so. Rather, the demise of deliberate weather and climate modification appears linked to the growth of environmental concerns' (Kwa in Miller and Edwards (2001: 152)).
28 Kellogg and Schneider (1974).
29 Kellogg and Schneider (1974: 1163).
30 Weart (2008: 19–38).
31 For a similar argument, see Baker (2017: 19–21).
32 Baker (2017: 19).
33 Turner and Isenberg (2018: 33). Of course, not all scientists agreed with this new role. Some climate scientists tried to actively counteract the environmental movement in an effort to safeguard their position within the state (see, particularly, McCright and Dunlap (2003); Lahsen (2008)).

34 Marchetti (1977).
35 Intergovernmental Panel on Climate Change, United Nations Environment Program, and Titus (1990).
36 Intergovernmental Panel on Climate Change, United Nations Environment Program, and Titus (1990: 108); US Senate, Committee on Environment and Public Works (1997: 131).
37 US Senate, Committee on Environment and Public Works (1997: 3, 13, 15, 24, 26, etc.).
38 US Senate, Committee on Environment and Public Works (1997: 1f.).
39 Kellogg and Schneider (1974: 1163).
40 Schneider in US Senate, Committee on Environment and Public Works (1997: 130).
41 US National Academy of Sciences (1992).
42 US Senate, Committee on Environment and Public Works (1997: 130).
43 See, e.g., Weart (2008: 139f.).
44 Taylor and Buttel (1992: 405).
45 Pielke (2000a: 16).
46 Pielke (2000a: 16).
47 For an instructive account of the career of biodiversity loss, see Hannigan (2006).
48 Wilson (2006); Hannigan (2006: 128).
49 US National Academy of Sciences (1977: 7).
50 Pielke (2000a: 11–12).
51 Keller (2009: 102).
52 Pielke (2000a: 12).
53 Pielke (2000a: 14). See also Fleagle (1986: 56).
54 Pielke (2000a: 17).
55 Pielke (2000a: 17).
56 Pielke (2000a: 19).
57 Pielke (2000a: 23).
58 Pielke (2000b: 136).
59 Pielke (2000b: 136).
60 Pielke (2000a: 10).
61 This rising political interest in climate change of course entailed structural dynamics within science as well. As Spencer Weart observed: 'Specialists in the quirks of the stratosphere, volcanoes, ocean chemistry, ecosystems, and so forth found themselves sharing the same funding agencies, institutions, and even university buildings' (2008: 144).
62 Pielke (2000a: 13).
63 Intergovernmental Panel on Climate Change (2017).

64 Fleming (2010: 226).
65 Nierenberg, Tschinkel, and Tschinkel (2010: 219); see also Pielke (2000a: 13). This also explains why the Department of Energy emerged as one of the most relevant organisations, providing climate change expertise within US policy during the late 1970s.
66 Nierenberg, Tschinkel, and Tschinkel (2010: 219).
67 See, particularly, MacDonald and others (1979); US National Research Council (1979); US National Research Council (1982, 1983); US National Academies of Sciences (1992).
68 Schelling in US National Research Council (1983: 449).
69 US National Academy of Sciences (1977: ix).
70 Andrews (2006: 223).
71 US National Academy of Sciences (1977: ix, emphasis in original).
72 For the suggestion of fertilising the oceans, see, e.g., US National Academy of Sciences (1977: 6).
73 US National Academy of Sciences (1977: 13).
74 US National Research Council (1982: xv). See also Nierenberg, Tschinkel, and Tschinkel (2010: 319).
75 US National Research Council (1983: 58–59).
76 US National Academy of Sciences (1992: 28).
77 The featured papers were, for example, Marland (1996); Schneider (1996); Bodansky (1996). According to Oldham and others, the publication of this issue sparked 'a first significant spike' in publication activity on climate engineering (Oldham and others 2014: 149).
78 See, e.g., Stilgoe (2015: 162).
79 Budyko (1977). See also Fleming (2010: 241).
80 Marchetti (1977). See also Schneider (1996: 292).
81 IPCC, United Nations Environment Program, and Titus (1990).
82 Marchetti (1977: 59).
83 See, e.g., Lawrence and Crutzen in Blackstock and Low (2019: 90).
84 US National Research Council (1983: 449).
85 Schelling in US National Research Council (1983: 469).
86 Schelling in US National Research Council (1983: 469).
87 US National Academy of Sciences (1992: 460).
88 'To date, the policy cart has been careering far in front of the scientific horse. Presidents convene climate conclaves. Prime ministers declaim on the need to reduce CO_2 emissions. Even a distinguished international panel of scientists, who should know better, calls for a 60% cut in these emissions. Yet, like most declarations of war, these calls to arms against global warming have been made without an attempt to weigh the costs and benefits of restraints' (Nordhaus (1990)).

89 Nordhaus (1990: 19). See also Nordhaus (1992: 1317).
90 Schelling (1996).
91 Schelling (1996: 306).
92 Schelling (1996: 303).
93 Schelling (1996: 306).
94 Turner and Isenberg (2018).
95 Turner and Isenberg (2018: 8); see also Jacques, Dunlap, and Freeman (2008).
96 Turner and Isenberg (2018: 17).
97 Turner and Isenberg (2018: 157).
98 See, e.g., McCright and Dunlap (2003).

CONCLUSION TO PART II

This second part of the book has illustrated that the recent rise of climate engineering as a 'bad idea whose time has come' is only the latest episode in a much longer standing history of ideas about deliberately modifying the climate. Chapters 3 and 4 have traced the historical roots of the current debate all the way back to the turn of twentieth century. This long-range perspective confronts a common assumption. It questions the often-propagated fracture that the daunting proposition of climate engineering has implied for climate policy agendas. This long-range perspective sheds light on another temporality of this debate, adding another piece of the puzzle, and thus providing a more differentiated picture of this career of climate engineering.

The early history of climate engineering illustrates that, despite being framed as a 'last resort' or 'Plan B', suggestions of techno-scientific climate intervention have been an essential part of the political cultivation of climate science from the outset. From the angle of these historical perspectives, it is not the notion of climate engineering which imposed a fracture on established science-politics alliances around climatic change. Instead, it was the politicisation of climate change which fractured historically grown alliances around promises of climatological control. In a way, this career of climate engineering is thus a particular history of the career of climate science in politics. The concept of targeted climate intervention has not been devised in responding to climate change. It has not been devised as the result of a linear innovation or science policy process. The discovery of the problem did not stimulate exploration of this response. Instead, problem and response have co-evolved from the very beginning; their histories coalesced. Climate engineering, then, can be understood as a node that has managed to effectively match scientific to political agendas around issues from agricultural interests to military interests, and eventually to environmental concerns. Each stage of the career of climate engineering has been defined by shifting alliances between politicians and climate scientists as they sought to advance their respective agendas.

Chapter 3 suggested that hopes for the positive prospects of deliberate climate control preceded fears of global warming. Notions of targeted climate intervention were rooted in the very first scientific explorations of human agency in climatic change around 1900. These observations were born out of a deeply divided and pre-disciplinary field of climatology. Descriptive meteorological efforts were institutionalised in a growing network of state-funded weather services, while central insights regarding physical theories of climate change were rather dispersed and brought forth by isolated and individual scientific efforts. The geopolitical challenges of the early- to mid-twentieth century then turned climatology into a project of national security, prestige, and progress. Meteorology became politically relevant, resulting in vast institutional and professional expansion of climate research. This massive infrastructure of climate research made the atmosphere into a subject that could not only be qualitatively described and mapped, but also 'rendered calculable' and – this was hoped at least – controllable.[1] Significant meteorological progress in observational and modelling capacities during the second half of the twentieth century further fuelled political visions of climate control.

Chapter 4 then traced how the politicisation of global warming between the 1970s and 1990s implied a substantial fracture for these hopes of climate control and the grown alliances between climate science and the state more generally. What had previously been primarily addressed as two sides of the same coin (inadvertent and deliberate modification), now appeared as a problem (anthropogenic climate change) and – cautiously and slowly – a potential response (techno-scientific climate intervention).[2]

In the first part of the chapter, we saw how political interest in climate and weather modification was drowned out precisely as climate change became politicised as a problem of global societal significance. Nurtured in the newly established institutions of environmentalism, climate change became problematised as a concern for environmental safeguarding; it rested on ecological observations on the fragility of nature. This meant that climate change became understood as a challenge to reduce rather than expand techno-scientific

CONCLUSION TO PART II

intervention capacities. It became understood as a problem that marked the limits of human control over the climate, one that highlighted the 'limits of growth', and that questioned the political and economic status quo. This had two relevant consequences for the career of climate engineering. On the one hand, this problematisation of climate change effectively curbed the earlier political excitement over the prospect of deliberate climate modification that had defined the 1930s through 1960s. On the other hand, this politicisation of climate change fractured established alliances between climate science and the state more broadly. Climate science no longer seemed to provide a tool of control at the hands of the state, but, quite to the contrary, seemed to be part of an agenda that questioned the economic and political status quo.

The politicisation of climate change not only questioned established alliances between climate science and the state, but it also forged powerful new ones. The second part of Chapter 4 suggested how, between the 1970s and 1990s, climate science became established as the *problem-defining authority* of this newly politicised issue. And although not pursued as a Plan A, we saw how climate engineering was couched in the very formulation of this 'greenhouse problem'. The chapter traced how these measures now emerged as catering to different visions of tackling this newly assembled challenge, and, by extension, to different visions of the role of climate science in the state. This second part of the chapter thus describes the roots of what might be described as an incremental re-normalisation of climate engineering in the context of climate change policy. From here on out, these measures would embark on a journey back into the limelight of US climate policy.

This second part of the book thus illustrated how the formulation of this 'bad idea whose time has come' bundles the disparate histories of various technological concepts – sprouting in different contexts and times – as well as the dynamic problematisation of climatic change as an agricultural, military, and environmental challenge. This perspective suggests that we are not, in fact, at a point zero today. Instead, we have systematically paved a way to arrive here. The forces which shape these emerging politics of climate engineering are grown forces of science in politics. This part of the book suggested how visions of techno-scientific climate control have successfully linked scientific to political agendas throughout different historical contexts, couched in the shifting issues and problems of their time. By considering these multiple threads of the story, it becomes clear that climate engineering emerges on policy agendas not merely because of the somehow external urgency of the issue at hand. Instead, this envisioned response measure is rooted in historically contingent modes of defining the problem.

NOTES

1 Baker (2017: 11).
2 Of course, this shift was incremental. Even in 1992, the National Academies still refer to anthropogenic climate change as 'inadvertent geoengineering' (US National Academy of Sciences (1992: 433)).

PART III

ENGINEERING THE CLIMATE

SCALING THE ISSUE AND SUGGESTING CONTROL

5
ASSEMBLING AN ENGINEERING PROBLEM

> 'The trouble with the global warming debate is that it has become a moral crusade when it's really an engineering problem.' Solving an engineering problem requires defining the goal quantitatively, facing the technical challenges, and creating systems to address these as cost-effectively as possible.
>
> Martin Hoffert before the House Committee on Science, September 20, 2006, quoting Robert Samuelson.[1]

WITHIN A MATTER OF A DECADE, THE CONGRESSIONAL DEBATES AND POLICY disputes surrounding the exploration of climate engineering in the United States changed substantially, not only in their tone but in their very objective. In this chapter, we will see how, beginning in early 2000, notions of climate engineering (re-) gained political currency in the United States when the very problem that these concepts promised to address was reformulated – that is, when climate change became assembled as a technological innovation challenge, when it was formulated as a project which would lend itself to techno-scientific intervention and control.

In contrast to the 1990s, climate change was no longer discussed as a curious scientific puzzle; instead, it emerged now as an urgent societal challenge – a challenge to be tackled, combatted, and 'won' by techno-scientific innovation. Put differently, it took transforming the problem that climate engineering promised to address to bring these measures further into the political limelight again. In response to this understanding of climate change as a challenge of technological innovation, notions of climate engineering moved from the footnotes of academic debates where they were found during the 1990s to the very core of controversial legislative inquires and executive efforts as we move into the early 2000s.

Reformulating climate change in this manner implied a new vision for climate science's relationship to the state. These shifting configurations in the alliances between climate science and the state once again defined this particular historical setting in the career of the climate engineering. From the 1970s to the 1990s, it had been outspoken scientists and social movements that shaped the politicisation of climate change, as we have seen in the previous chapter. These groups had emphasised environmental safeguarding and formulated climate change as a challenge involving a reduction, rather than an expansion, of techno-scientific intervention capacities. During the first decade of the new millennium, the political exploration of climate engineering suggested the antithesis to this scenario. Climate engineering was now pushed onto the congressional agenda in the context of calls to properly harness climate science as a tool for the state in addressing the climate change issue. Climate science was no longer envisioned as questioning the political and economic status quo, but instead emerged as a potential tool for stabilising it. Climate engineering gained further traction, in other words, as climate science evolved from *problem-defining* to a *problem-addressing* authority. In a sense, this emerging role of climate science in the state thus realised a vision that had been looming already in the economic analyses which we explored in the final section of the previous chapter.

In this third part of the book, we will trace a kind of renaissance of climate engineering measures in US politics. In this kind of re-kindled vision of techno-scientific control of the climate, climate engineering mobilised otherwise conflicting (even competing) constituencies over the issue of climatic change. It engaged Democrats and Republicans, conservatives and progressives, climate scientists and oil companies in rather controversial policy disputes and legislative inquiries over what was at stake.

This and the following chapter trace this re-invigorated debate over climate engineering through two very different political landscapes, respectively coinciding roughly with the Bush and Obama presidencies. We will begin in this chapter by exploring the status of climate engineering prior to the official inquiry into these measures from 2009, focusing in particular on the years from 2003 to 2007. The chapter begins by examining the specific political problematisation of climate change that defined the debate over climate engineering during these years. We will also take a closer look at the kinds of expert observations that were essential for assembling climate change in this context. The chapter then follows climate engineering through the policy process and explores how it took political shape during the early 2000s. In doing so, it focuses on three distinct contexts: how climate engineering was discussed in legislative disputes over the status of science and technology within US climate policy, how it became subject to highly controversial expert testimonies, and finally, how it gained political traction by mobilising an economic and managerial gaze onto the issue of climate change. Each of these three contexts will serve to further substantiate how the career of climate engineering corresponded to the aforementioned shifting alliances in climate science and the state as well as a new and rather diverse constituency for climate policy matters.

DO WE NEED A MANHATTAN PROJECT FOR THE ENVIRONMENT?

To make sense of how the politics of climate engineering have evolved thus far in the twenty-first century, we need to consider the period's particular political landscapes. As we will see in this and the following chapter, the political exploration of climate engineering seems somewhat formatted by a shift in administration. The presidency of George W. Bush, lasting from 2001 until 2009, provided the defining political environment for the career of climate engineering in the period covered in this chapter, while the incoming Obama administration, lasting from 2009 to 2017, provided the relevant context for the period covered in the next chapter.

In 2001, George W. Bush won a highly controversial presidential election against Democratic candidate, Al Gore. While climate change had played virtually no role in either of the 2000 presidential campaigns, the election of Bush drastically intensified politicisation of the issue, effectively driving it to the heart of partisan politics.[2] Beginning in the early 2000s, we can trace a general invasion of US American politics by 'the specter of abrupt climate change'.[3] This was driven by increasing public concern for the climate change issue in the United States – having reached a 'historic high' in 2000.[4] In the years following the election, climate change thus became the subject of fierce political dispute and controversy. In 2006, in the middle of George W. Bush's second term as president, Al Gore released the high-profile film, 'An Inconvenient Truth'. It formed a key part of Gore's extensive campaign to educate the American public about the problem of global warming. Almost simultaneously, Arnold Schwarzenegger signed into law the first cap on CO_2 emissions within the United States as Governor of California.[5] The Bush administration, on the other hand, followed a strategy of halting and forestalling policy action on climate change and dismantling regulatory initiatives.[6] By the first half of 2007, partisan conflict and public interest in the issue had driven climate change onto the agenda of 39 congressional hearings.[7]

As the issue of climate change became the subject of a fierce political and partisan conflict, the role of science and technology in fighting the issue emerged as a core battleground.[8] The political problematisation of climate change during these years was marked by an increasing sense of urgency and intensified politicisation of climate change, while, at the same time, re-invoking hopes of techno-scientific control as a means for addressing this issue.

We can get a sense of this notable shift in how climate change became assembled during this period by zooming into some of the expert observations that defined congressional debates at the height of these conflicts. In September 2006, Martin Hoffert, emeritus professor for physics at New York University opened his testimony before the House Committee on Government Reform by quoting John F. Kennedy:

> We choose to go to the moon in this decade and do the other things, not because they are easy, but because they are hard, because that goal will serve to organize and measure the best of our energies and skills, because the challenge is one that we are willing to accept, one we are unwilling to postpone, and one we intend to win.[9]

We met Hoffert at the beginning of this chapter – as the author of the introductory quote taken from the very same testimony, in which he suggested that climate change needs to be understood as an engineering challenge. Hoffert's testimony provides an instructive starting point into these controversial discussions, as it reflects and very much pinpoints the new gaze onto the climate change issue that accompanied the political exploration of climate engineering during these years. Hoffert was invited to testify – and he ultimately did so across two separate hearings – in the context of an inquiry into the status of the Bush administration's efforts to foster climate change technology research. The title of one of the hearings asked, 'Do we Need a Manhattan Project for the Environment'? I will unpack the status and details of this hearing in more detail later during this chapter. For now, I simply want to focus on the programmatic status of this question for the problematisation of climate change during the early 2000s and the implied role of science and technology in addressing the problem.[10]

Beginning in 2003 and particularly around 2006, climate policy actors and experts alike began invoking a set of high-profile techno-scientific projects that had forged a tight bond between science and the state during the Second World War and in its aftermath. The first was the Manhattan Project. Between 1939 and 1946, the Manhattan Project had resulted in the development of the first nuclear weapons. Now, in discussions of climate change, it was repeatedly mobilised to invoke visions of national strength in the face of this challenge. The analogy served to suggest the grandeur of the climate change challenge, all the while emphasising that this challenge, too, would similarly be manageable by techno-scientific innovation.[11] Other reference points during the period included Project Apollo, which had succeeded in putting a US American on the moon, and the establishment of military research and development infrastructure such as the Defense Advanced Research Projects Agency (DARPA).[12]

These comparisons channelled science and technology as a kind of weaponry – as a form of national strength and security in tackling the climate change challenge. In these observations, counteracting climate change is like building the atomic bomb or landing on the moon; it is a challenge that would take 'the greatest engineering effort in history [...]'.[13] 'What we are faced with', Martin Hoffert suggested, 'is a kind of existential challenge to our high-technology civilization'.[14] Climate change appears in these observations as a serious challenge, manageable only by a targeted and orchestrated national effort aimed at steering the right kind of research and development. It is presented as controllable, in other words, by the development of the right kinds of techno-scientific weaponry.

Quantified modes of observing were essential in assembling this gaze onto the issue of climate change. This quantified gaze served to scale the issue of climate change; it served to suggest both the grandeur of the challenge as well as its techno-scientific manageability. Around 2006, numeric observations of climatic change became increasingly standardised, with experts and policy actors alike increasingly invoking the symbolic power of distinct numbers. Take, for example, the target of limiting global warming to 2°C compared to pre-industrial levels, along with the goal of stabilising concentrations of atmospheric CO_2 at 450 parts per million (ppm). Both goals made their first appearance in the examined policy documents in 2006. And these distinct numbers were used to argue for the need for more concrete policy goals in tackling the climate change challenge. For instance, Hoffert mobilised the 2°C temperature target in a critique of the Climate Change Technology Program:[15]

> This is the real problem. The Manhattan Project didn't aim to explore nuclear weapons in general; its goal was building a Bomb before the end of WWII. The Apollo Program didn't aim at exploring manned space flight in general; it's goal was putting a (US) man on the Moon by the end of the '60s. So too does the CCTP [Climate Change Technology] program need a more concrete goal [...] Tony Blair at the recent Exeter conference in the UK set an upper limit of two degrees Celsius global warming.[16]

By proclaiming climatic thresholds, targets, and ecological tipping points, these observations quantified 'the size of the world's job'.[17] Setting the right kind of goal in this sense invokes a sense of control in the face of this challenge. It appeals to urgency and the manageability of climate change at the same time. On the one hand, these 'numbers' provided the formerly diffuse atmospheric and oceanographic phenomenon with tangibility. By invoking the daunting environmental catastrophe, they reinforced the urgent need to tackle this challenge.[18] On the other hand, these targets and thresholds transformed climate change into a manageable challenge, a challenge that seemed subjectable to economic logic and techno-scientific management, even control. Climate change becomes an 'engineering problem', as Hoffert suggested at the outset of this section; it becomes a challenge that 'requires defining the goal quantitatively' to then 'fac[e] the technical challenges', and eventually 'creat[e] systems to address these as cost-effectively as possible'.[19]

Social scientific scholarship has suggested how such thresholds and targets combine scientific observations with policy directives.[20] Bettina Heintz, for example, called attention to the dual nature of numerical observations as both representing and making reality.[21] Numerical observation emerges here as a means of 'world-creation' (*Welterzeugung*).[22] Theodore Porter prominently coined the expression of 'speaking precision to power' in this context.[23] This means that although climate thresholds or tipping points might not necessarily be accurate, they successfully invoke a sense of precise and thus legitimate observation. Therefore, this 'indexed language' of climatic targets offers the suggestion of control and thereby reinstates the political capacity to act in an otherwise hopelessly complex situation.[24] It seems fitting in this sense that it was mostly policy actors (and not, for example, scientific experts), who called for the quantification of distinct stabilisation targets to prevent dangerous climate change.[25] As Senator Waxman put it, '[i]f we don't pick a goal and the right goal, we may be aiming for disaster'.[26]

It was this hope for techno-scientific control in the face of a national challenge that brought a White House which had been 'officially sceptical' of global warming to start to explore climate intervention measures.[27] It was this vision of techno-scientific control that provided the relevant context and defining breeding ground for the highly controversial inquiries into climate engineering during these years. Climate science appeared not merely as an academic endeavour in this context but prevailed as a powerful political – and particularly national strategic – force. It was mobilised as the very tool that would provide the nation with agency in tackling the issue of climate change. Against this backdrop, we will now explore three defining contexts in which climate engineering took political shape in the period between 2000 and 2009. We will see how climate engineering became subject to controversial legislative disputes over the status of science and technology within US climate policy, how it was visibly pushed onto the congressional agenda in the context of highly controversial expert observations on these measures, and finally, how it gained political traction by speaking to economic and corporate concerns regarding climate change.

'FROM DEBATING SCIENCE TO FINDING SOLUTIONS': CLIMATE SCIENCE AS A TOOL FOR THE STATE

In September of 2006, Tom Davis was one of the first[28] members of Congress to bring the issue of climate engineering into congressional debate, when he criticised the Bush administration's lack of attention to these measures. 'The federal government', he complained, had not yet engaged 'in any exploratory or innovative technology research on climate change', leaving 'climate clinicians [sic] that lie outside of existing technology, such as geo-engineering and artificial photosynthesis [...] unaddressed'.[29] Tom Davis, former Republican member of the US House of Representatives, was speaking here as Chairman of the Government Reform Committee, one of the most powerful congressional committees, responsible for government oversight. He was opening the very hearing which we touched upon earlier in this chapter, assessing US 'Climate Change Technology Research' and asking whether the United States would need a 'Manhattan Project for the Environment'. Davis' concern for the need for climate engineering research was couched here into a new vision for climate science in the state.

This hearing as well as Davis' critique of the lack of federal attention to climate engineering measures was part of an ongoing legislative inquiry into the Bush administration's climate change technology initiatives during the early 2000s.[30] In this section, we will see how climate engineering moved further into the political limelight in the context of a bigger conversation regarding the status and promise of technological innovation as a national approach to climate change. In a number of hearings between 2003 and 2007, policy actors and experts controversially discussed the place of climate engineering research within such a national approach – particularly in the context of assessing the

Bush administration's Strategic Plan for its so-called Climate Change Technology Program (CCTP), which had just been released in September of 2006.

The Climate Change Technology Program was established in 2002, as part of the Bush administration's effort to expand the 'technology component' of US climate change research.[31] To this end, Bush launched the CCTP as a multi-agency initiative, led by the Department of Energy, and relabelled the existing US Global Change Research Program as the Climate Change Science Program (CCSP).[32] In this new set-up, the two programs, CCTP and CCSP, were intended to differentiate and strengthen the 'technology' component of US climate change research from the 'science' component.[33] These programs thus reflect the shifting status and role of climatological expertise within the state during these years. They are the product of an administration that, on the one hand, was averse to policy action against climate change, yet, on the other, felt the pressure of rising public concern over the issue.[34]

In 2006, the CCTP finally released its much-anticipated Strategic Plan. The plan attributed around 3 billion dollars in federal spending for 'climate technology research, development, demonstration, and deployment' and presented a 'planning-horizon' for no less than 100 years.[35] Former Secretary of Energy, Samuel W. Bodman described the document as '[...] inspired by the President's vision to harness America's strengths in innovation and technology' to provide a more sustainable energy system.[36] Climate change research was invoked here as the solution itself: instead of questioning the economic and political status quo, it appeared as a national tool to trigger technological innovation in tackling climate change.

This new vision for climate change research within the state was precisely the defining context which brought Congressman Davis to call for the need for climate engineering research. When the CCTP presented this Strategic Plan to Congress in 2006, it came under harsh scrutiny, triggering, among other concerns, controversial debates over the need of climate engineering research to become part of the program. Notably, this was despite the fact that the plan did include some forms of climate engineering measures – specifically terrestrial and ocean sequestration measures.[37] Davis addressed his fellow members of Congress in the House Committee on Government Reform just one day after the plan's publication:

Good morning, and welcome to today's hearing on climate change technology. As we sit here today, the debate over climate change science continues, but this Committee [on Government Reform] [...] – as well as the Administration and many others in government – have already recognized the important facts: that global mean temperature has increased over the past century, and that carbon dioxide in the atmosphere has contributed in some way to this warming. *With this in mind, our committee seeks to move away from debating science to finding solutions.*[38]

Davis' observations thus nicely pinpoint the shift that we have traced regarding the status of climate science in the state from the previous chapter to this one: in the 1970s through 1990s, climate science had become established as the authority to 'discover', assemble, and define the carbon dioxide problem (see Chapter 4). This had stoked debates over the basic epistemological premises of the climate change issue. Climate change, to use Davis' words, became primarily subject to debates over 'the science' of this issue. In contrast to this outlook, this legislative inquiry during the early 2000s now advanced a vision of climate science – and particularly technology – as a central political asset in addressing the climate change issue. Climate science thus shifted its status from a problem-defining to a problem-addressing authority in these debates; it appeared as literal weaponry at the service of government. The committee, Davis announced, 'has taken an important step by discussing how the Federal Government can better arm itself with technology to address this worldwide problem [of climate change]'.[39]

Following this line of reasoning, one central issue on the committee's agenda was the potential and promise to develop an 'ARPA for climate change'.[40] ARPA in this case stands for Advanced Research Projects Agency and refers to an agency that President Eisenhower had established during the Cold War years in response to the Soviet launch of the world's first artificial satellite, Sputnik. During the 1970s, ARPA was renamed DARPA, the Defense Advanced Research Projects Agency. The organisation still exists today. ARPA essentially implies an organisational model specifically devised to foster technological breakthroughs, a 'central, authorized body to command exploratory research'.[41] The idea was to create an agency within the federal government that would focus on 'high-risk, high payoff' research in a relatively independent, 'non-bureaucratic' setting with little oversight and an emphasis on exploratory research – i.e. 'risk-taking and tolerant of failure, open to learning'.[42] By suggesting this organisational model for climate change research, the congressional debate thus further specified the new problem-addressing vision that emerged during these years for climate science in the state. These observations suggest organising climate research in a manner that would turn basic science into 'solutions' – much like DARPA, which, according to Congressman Davis, 'was created to turn innovative technology into military capabilities' and by doing so, 'produced not only military advancement but commercial benefits, as well'.[43] The DARPA model, in other words, promised an organisational blueprint for generating, even systematically programming, politically relevant research; it promised a set-up that would seamlessly match scientific to political interests.

This discussion over a new organisational model for triggering climate change technology stood in the context of a bigger inquiry, spearheaded by members of both parties, who, in 2005, asked the US National Academies of Sciences how the United States could 'maintain leadership in key areas of science and technology'.[44] In their report, which was officially published only in 2007, the academies suggested establishing the ARPA model for energy research – an ARPA-E within the Department of Energy. While Bush signed ARPA-E into law in 2007,[45] just one year after the House Committee on Government Reform hearing, an ARPA for climate change is yet to be realised. That said, the idea was brought back onto the agenda in 2020 as part of the presidential campaign by the Biden administration.[46]

Climate engineering appeared in these debates during the early 2000s as a highly controversial example for the kinds of 'exploratory' or 'high-risk' research that would be implied by such a new approach to federal climate research. Expert witnesses advocated both for and against the need to include climate engineering research into such a federal program. Lee Lane, for example, fellow at the American Enterprise Institute, a conservative policy think tank, appeared as a vocal advocate for climate engineering research. Lane strongly advocated for the urgent need to 'expand the program's [CCTP] agenda to include geoengineering' in his testimony.⁴⁷ According to Lane, the option to engineer the climate would provide an 'insurance against runaway climate change' that could, in contrast to the regulation of emissions, be 'implemented swiftly'.⁴⁸ Richard Van Atta from the Institute for Defense Analyses, in contrast, drew on his experience with the DARPA initiative to voice concerns regarding an ARPA model as the fitting institutional setting to oversee and direct research like 'dispersing particles in the atmosphere'.⁴⁹

PUSHING A CONTROVERSIAL FIX: EXPERT WITNESSES AS AGENDA SETTERS

A second defining context, in which climate engineering took political shape in the first decade of the 2000s, was a set of dispersed and highly controversial congressional expert testimonies. In stark contrast to the cautious, indirect, and rather well-hidden congressional references to climate engineering during the 1990s, these testimonies were now loud and clear – at least regarding their

NAME	INSTITUTIONAL AFFILIATION	TYPE OF INSTITUTION	HEARINGS IN WHICH THE EXPERTS TESTIFIED	YEAR
Schnare David	Thomas Jefferson Institute for Public Policy	Think Tank	The Impacts of Global Warming on the Chesapeake Bay	2007
Lane Lee	Climate Policy Center	Think Tank	Climate Change Technology Research: Do We Need a 'Manhattan Project' for the Environment Geoengineering: Parts I, II, and III	2006, 2009
Doney Scott	Woods Hole Oceanographic Institution	Research Institute	Effects of Climate Change and Ocean Acidification on Living Marine Organisms	2007
Romm Joseph	Center for American Progress	Think Tank	Voluntary Carbon Offsets – Getting What You Pay For	2007
George Russ	Planktos Inc.	Corporation	Voluntary Carbon Offsets – Getting What You Pay For	2007
Feely Richard	National Oceanic and Atmospheric Administration	Federal Agency	Effects of Climate Change and Ocean Acidification on Living Marine Organisms	2007

ASSEMBLING AN ENGINEERING PROBLEM

NAME	INSTITUTIONAL AFFILIATION	TYPE OF INSTITUTION	HEARINGS IN WHICH THE EXPERTS TESTIFIED	YEAR
Green Kenneth	American Enterprise Institute	Think Tank	Drought, Flooding and Refugees: Addressing the Impacts of Climate Change in the World's Most Vulnerable Nations Building US Resilience to Global Warming Impacts Combating Climate Change in Africa Not Going Away: America's Energy Security, Jobs and Climate Challenges	2009, 2010
Conover David	Stoney Brook University	University	Effects of Climate Change and Ocean Acidification on Living Marine Organisms	2007
Hoffert Martin	New York University	University	Department of Energy's Plan for Climate Change Technology Programs Climate Change Technology Research: Do We Need a 'Manhattan Project' for the Environment	2006
Eule Stephen	US Department of Energy	Governmental Department	Climate Change Technology Research: Do We Need a 'Manhattan Project' for the Environment	2006
Andrews Clinton	Rutgers University	University	Public Transportation: A Core Climate Solution	2009
Solomon Susan	National Oceanic and Atmospheric Administration	Federal Agency	Commerce, Justice, Science, and Related Agencies Appropriations for 2010	2009
Moniz Ernest	MIT	University	The Future of Coal Department of Energy: Science and Technology Priorities	2007, 2014
Van Atta Richard	Institute for Defense Analysis	Research Institute	Climate Change Technology Research: Do We Need a 'Manhattan Project' for the Environment	2006
Holdren John	Office of Science and Technology Policy	Executive Office of the President	Climate Services: Solutions from Commerce to Communities	2009
Haass Richard	Council on Foreign Relations	Think Tank	United States – China Relations in the Era of Globalizations	2008
Figdor Emily	Environment America	Think Tank	The Role of Offsets in Climate Legislation	2009
Deutch John	Massachusetts Institute of Technology (MIT)	University	The Future of Coal	2007

TABLE 6.1 Expert Witnesses Mentioning Climate Engineering (2006–2009)

own position on the issue. Experts and policymakers now started both pushing and challenging climate engineering as a 'fix' for climate change. They variously suggested it as a 'great'[50], 'inevitable'[51] 'potential'[52], 'risky'[53], 'unproven'[54], 'very wrong'[55], or simply 'bad'[56] approach for counteracting climate change. In this section, we will see how these dispersed and highly controversial testimonies provided a kind of odd, yet critical arena in pushing climate engineering further into the political limelight. These testimonies served as a kind of masked agenda setting context for climate engineering – masked, because they strongly advocated for or against these measures before the issue was officially introduced to the political agenda, that is, before policymakers took an official stance on the topic.

A case in point for such an agenda setting expert account was a 2007 testimony by David Schnare in a hearing on, 'The Impacts of Global Warming on the Chesapeake Bay', before the Senate Committee on Environment and Public Works.[57] At the time of his testimony, David Schnare was part of the Thomas Jefferson Institute, a non-for-profit think tank based in Virginia, which has since come under critique for its ties to climate denialist organisations.[58] Aside from his role at the Thomas Jefferson Institute, Schnare also has had a long career with the Environmental Protection Agency (EPA) – most recently in 2017, as part of the Trump administration's EPA transition team. In his testimony, he strongly advocated for the need for solar radiation management research and development. Schnare urged the committee to address climate engineering as a first response to climate change. In his opinion, global leadership regarding the responsible assessment and deployment of these measures remained an 'unmet national duty'.[59] He argued that 'absent some form of geo-engineering […] it is too late to prevent melting of the Greenland Ice Sheet, and the planet will suffer a 23-foot rise in ocean levels'.[60] In his testimony, we can trace the relevance of scaling the issue of climate change for advancing climate engineering as a critical response measure. Scaling this issue served to assess the appropriate response; it served to legitimise radical approaches and delegitimise others as 'pious':

> the question of incremental approaches crashes on the rocks of the time scales with which we are operating. If we are to prevent 550 parts per million of CO2 in our atmosphere, which is considered the point at which we

hit the first tipping point, the inevitable full melting of the Greenland ice sheet, some argue, including Nobel laureate Paul Crutzen, that it is already too late, and that any attempt to prevent that is nothing more than, in his words, 'a pious hope'.[61]

David Schnare not only introduced but visibly marked climate engineering into the political record by adding several papers on geoengineering in full to the congressional record.[62] He managed to fill 50 some pages with numerous references to likeminded scholars and scientific studies at a time when climate engineering was only beginning to be explicitly picked up in congressional debate.[63]

Similarly positive were the accounts of expert witnesses such as Lee Lane and Kenneth Green, the latter of whom criticised the focus on mitigating greenhouse gas emissions as 'misplaced'.[64] Kenneth Green pushed the topic in a notably expansive and repetitive intervention, by advocating for the need to invest in climate engineering research and development across four different hearings in almost identical testimonies.[65] David Conover, Scott Doney, Richard Feely and Russ George, provided less enthusiastic accounts in their explorations of the potential of ocean fertilisation measures,[66] and Obama's science advisor, John Holdren, rather played down the relevance of climate engineering to the incoming Obama administration when asked about the topic.[67] Other expert witnesses were highly critical of the potential of technological climate intervention. Emily Figdor from Environment America – an environmental protection advocacy group – for example, strongly dismissed the viability of ocean fertilisation measures as part of US climate legislation, when discussing potential carbon offset projects.[68]

Despite their outspoken positions on the issue, the expert witnesses who introduced climate engineering to the congressional agenda here, hardly appeared as prominent experts on the matter – at least as indicated by the US political inquiry into the issue: these witnesses made their statements on climate engineering in the context of thematically diverse hearings, spread over several years, and rather isolated from any systematic debate of the issue. Most of these experts were not invited to be part of the programmatic congressional inquiry into climate engineering that would begin in November of 2009. And what is more, they were not prominently referenced or referred to in the context of this systematic inquiry. In fact, twelve of the eighteen experts that raised the issue of climate engineering between 2000 and 2009 did not appear in the examined policy documents on the issue every again.

Seen individually, these experts thus appear rather irrelevant in this context of the politics of climate engineering. When taken as a group, however, they can be seen as playing a key role in shaping the career of these measures in US politics. These experts swiftly introduced a controversial issue into the congressional debate – apparently without the political initiative of congressional representatives themselves. As we have seen in Chapter 2, hearings provide Congress with essential leeway to place and navigate issues on the US political agenda and generate an evidence base for crafting legislation.[69] Depending on Congress's specific partisan composition, it can utilise these hearings either to support or to challenge the executive branch or the current administration. Aside from their role in sourcing information, hearings are thus essentially about asserting and contesting controversial issues and shaping the context for their political assessment.

Against this backdrop, these expert witnesses can be understood as playing the role of masked agenda setters. Their inputs allow policymakers to introduce a controversial issue onto the agenda without having to yet take an official stance on the topic. And in so doing, these experts pave the way for a programmatic congressional assessment of the issue at stake at a later point in time, allowing policymakers to refer back to their testimonies. In the case of David Schnare's testimony, for example, the hosting Chairman explicitly urged Schnare to provide a policy framework on climate engineering:

> Dr. Schnare, thank you very much for bringing the geo-engineering information. We will come back to it in time, but I would just invite you, if you have a framework that you would like to bring to my committee's attention, we would welcome this, because I think it will be a topic that will move on the global screen. I have questions and yellow lights about it. But rather than us giving our opinions about it, let's go beyond opinion and go to sound data and research, which is what we have been talking about here today.[70]

As requested, Schnare provided such a framework titled 'To Prevent the Catastrophic Effects of Global Warming Using Solar Radiation Management (Geo-Engineering)'. The Committee then added it to the record together with his prepared testimony.[71] In effect, this means that without the committee having to formulate an official position on this controversial matter, climate engineering was placed 'on the record'. By doing so, Congress can build on already established expertise on climate engineering – on 'sound data and research' – without having to establish an official inquiry into the issue. The selection of congressional expert witnesses is a highly strategic and purposive part of legislative activity in this sense.[72] Expert witnesses are expected '[...] to play a role in meeting the goals the chair has for the hearing'.[73] They provide a kind of 'ideologically and politically reliable' form of expertise, a type of 'staged advice', as we have seen in Chapter 2.[74]

The role of these expert witnesses in bringing the issue of climate engineering onto the congressional agenda also speaks to political science scholarship which has emphasised the shifting role of scientific expertise across different phases or stages of the policy process. Keller, for example, finds that scientific experts are more likely to provide explicit advocacy during the agenda setting phase, while during later stages of the policy process, the provision of expertise becomes increasingly formalised.[75]

GETTING WHAT YOU PAY FOR: THE 'INCREDIBLE ECONOMICS' OF CLIMATE ENGINEERING

Economic and corporate efforts to mobilise visions of techno-scientific control over the issue of climate change provide a third and final defining context in which climate engineering appeared in US climate policy during the early 2000s. This context further substantiates the emerging shift in the status of climate science from a problem-defining to a problem-addressing authority during this particular stage in the career of climate engineering. The political exploration of climate engineering was now increasingly built on the kinds of economic observations that had advanced climate engineering measures in the context of scientific assessments of climate policy since the 1980s and 1990s (see Chapter

4). Climate science in this context was not only envisioned as a tool for the state, as weaponry for the nation, but also as a potential economic asset, an investment opportunity. It was seen as offering, in other words, a managerial gaze onto the issue of climate change.[76] This was a new outlook that did not mark the limits to growth or question the economic status quo, but one that would promise control and provide business opportunities or economic solutions.

On the surface, this outlook became most obviously visible in some of the expert testimonies of the time. The accounts of Lee Lane and David Schnare are particularly illustrative in this context, as well as the papers by Alan Carlin and Scott Barrett, which were part of the group added by Schnare to the congressional record, as mentioned earlier.[77] In addition, the economic analyses from the 1980s and 1990s which we encountered in the previous chapter became increasingly prominent reference points within congressional debates of the early 2000s. Lane and Schnare, for example, referred in their testimonies to the analyses by William Nordhaus, Thomas Schelling, and the 1992 report by the National Academies of Sciences, which we explored in the final section of the previous chapter.[78]

Each of these accounts sought to establish the viability of targeted climate intervention in economic terms – as 'surprisingly cheap',[79] economically 'incredible'[80], or presenting a 'risky gamble'.[81] They formulated the goal of reducing emissions as 'well intentioned and even helpful', yet as 'inflexible, expensive, risky, and politically unrealistic' as a main policy strategy for tackling climate change.[82] Economic observations became essential in this context for advancing climate engineering in contrast to other policy measures. They provided the grounds for formulating these measures as superior to other mitigation options, even suggesting them as 'inevitable',[83] in contrast to other mitigation options:[84] 'keep in mind that use of geo-engineering will pay for itself, while exclusive reliance on greenhouse reduction will not only fail to pay for itself, it will fail to prevent global warming', as Schnare put it in his testimony.[85] Or: 'it is unlikely that cost would play any significant role in a decision to deploy stratospheric scatterers because the cost of any such system is trivial compared to the cost of other mitigation options', as Barrett suggested.[86]

Beyond this superficial layer of overly enthusiastic expert observations, economic and corporate concerns guided the set-up of a climate engineering-relevant research and development infrastructure. In the following two sections, we will see how at the turn of the new millennium notions of 'clean coal', carbon capture, and offsets began providing an important platform for advancing climate engineering research in the name of economic climate solutions.[87]

The rise of corporate and governmental research infrastructures

Part of the reason that climate engineering gained political currency during the early 2000s was that it mobilised what had thus far seemed like a rather unlikely constituency for the issue of climate change. This included corporate interests, particularly associated with the fossil fuel industry. In 2003, for example, the Federal Register gave notice that three of the world's largest energy and oil companies established a research and development project devoted to climate change and energy issues. Exxon Mobil Corporation, General Electric Company, and Schlumberger Technology Corporation initiated the Global Climate and Energy Project (GCEP), a commercially funded research and development initiative, located at Stanford University.[88] This project serves to illustrate how growing corporate interests in advancing technical fixes to tackle climate change aided in further establishing climate science as a problem-solving authority. Climate research was mobilised by the project not as 'raising the alarm' about an increasingly urgent problem,[89] but as providing 'new solutions to one of the grand challenges of this century'.[90] These 'new solutions' also included climate engineering-relevant research. With its focus on energy research, the project provided essential insights for advancing carbon dioxide removal (CDR) approaches. During its almost 17 years of operation, the Global Climate and Energy Project was dedicated to 'long-term pioneering research to identify options for commercially viable, technological systems for energy supply and use with substantially reduced net greenhouse emissions', including 'fundamental science and pre-commercial research' in 'carbon sinks, carbon dioxide separation and storage'.[91] The project ended in August of 2019.

A diverse set of international carbon capture and storage (CCS) projects, which began popping up in congressional debates during the early 2000s further substantiate this growing corporate interest in climate intervention measures. Members of Congress and expert witnesses began mobilising these projects to both confirm[92] and question[93] the technological readiness of climate engineering measures. This included references to large-scale geologic sequestration and ocean fertilisation projects such as the Weyburn Project (initiated in 2000 and steered by the Canadian Department of Natural Resources),[94] the In Salah Project (established in 2004 by BP, Sontrach and Statoil in Algeria),[95] and the Sleipner gas field in the North Sea.[96] To take the Sleipner gas field as an example, it contains the world's first industrial scale CO_2 storage unit and is the longest continuing CO_2 injection project initiated to date. It is operated by Equinor (formerly Statoil), Norway's state oil company. Equinor built the unit in 1996 to avoid paying CO_2 taxes on its natural gas production. At an offshore platform used to extract natural gas, CO_2 is simultaneously removed from the gas produced and then injected in the Utsira formation, a deep saline reservoir about one kilometre below the North Sea floor, off the shores of Norway. Since its inception, the Sleipner unit has led to the storage of over 16 million tons of CO_2 underground.[97]

The Executive branch, too, internalised this new managerial gaze onto the issue of climate change. Through its Office of Fossil Energy, the Department of Energy, for example, began advancing carbon capture and storage (CCS) research and development as early as 1997.[98] The Office's diverse set of Federal and private sector partners – ranging from the US Geologic Survey, the National Science Foundation (NSF), the US Department of Agriculture (USDA), the Department of the Interior (DOI), and the Environmental Protection Agency (EPA) to representatives from the oil industry[99] – suggests just how multi-faceted the political exploration of the topic has become since the 1990s.[100] The carbon capture and storage research portfolio of the department includes formats such as 'industry cost-shared technology development projects, university research grants, collaborative work with other national laboratories' as well as 'in-house' research through national laboratories.[101] Particularly through its so-called Carbon Sequestration Program, the Department of Energy has advanced climate engineering relevant research and development.[102] The program's goal was to 'clean up' fossil energy sources by first 'demonstrat[ing]' a series of safe and cost-effective technologies at a commercial scale' before 'establish[ing] the potential for deployment leading to substantial market acceptance'.[103] These 'safe and cost-effective technologies' included, among other approaches, ocean fertilisation measures.[104]

These corporate and federal initiatives illustrate how climate engineering relevant research became advanced during the early 2000s through the incremental development of a diverse industrial-scientific research and development infrastructure. They are realisations of emerging alliances between climate science and corporate interests forged by the promise of techno-scientific control. All of these initiatives were driven by the goal to provide economical 'solutions' to the climate change issue. And thereby, they importantly aided in institutionalising this new vision of climate science as a problem-solving authority in tackling climate change.

Carbon offsets: Capturing and storing a negative commodity

Finally, I want to turn to a controversial congressional debate over so-called carbon offsets that flared up in 2007. The debate reveals another facet of the economic problematisation of climate change and the corresponding new vision of the relationship between science in the state. Specifically, it suggests how this problematisation of climate change shaped not only the setup of a research and development infrastructure, but also legislative inquiry into climate engineering measures.

The basic idea of carbon offsets is to 'make up' for already emitted CO_2 by purchasing and trading titles to measures that reduce or remove atmospheric CO_2 emissions. Such carbon offsets thus differ from concepts like 'clean coal' or 'carbon capture and storage' in how the emission of carbon into the atmosphere and the capture of carbon from the atmosphere are integrated. Carbon capture and storage units integrate the removal of CO_2 and the generation of energy physically and structurally – the Sleipner unit, for example, removes CO_2 directly from its generated gas stream *before* the CO_2 would be emitted to the atmosphere and before the gas is pumped to the shore and distributed. Carbon offsets, in contrast, integrate the emission and capture of carbon indirectly, namely via offset markets. These markets develop criteria, measures, and certificates that verify and account for offsets and that are intended to determine the quality and value of different types of offsets.

In July 2007, the House of Representatives' Select Committee on Energy Independence and Global Warming held a hearing on the topic of 'Voluntary Carbon Offsets – Getting What You Pay For'. Among the invited witnesses was Russ George, CEO of Planktos Inc. Planktos was a private for-profit enterprise that sought to generate and sell carbon offsets by sequestering and storing CO_2 in the Pacific Ocean via ocean fertilisation. Earlier that year, the company had announced plans to seed a 100 km by 100 km area of the Pacific Ocean, close to the Galapagos Islands with approximately 100 tons of iron dust.[105] On 21 May 2007, the US Environmental Protection Agency (EPA) had contacted Planktos and required additional information regarding the venture's planned iron fertilisation project.[106] When Planktos was unable to provide that information, the United States submitted an agenda item to the Scientific Group of the London Convention on 1 June.[107] The London Convention is one of the first global conventions addressing pollution at sea. In their note to the scientific group, the United States expressed concern over the potential environmental impacts of Planktos' ocean fertilisation activities. According to the submitted agenda item, Planktos had informed the EPA in response to these concerns that it was no longer planning to conduct its ocean fertilisation activities from the Weatherbird II, a United States flagged vessel, but instead 'use a non-United States flagged vessel for releasing the iron so as not to be subject to regulation under the United States' Ocean Dumping Act'.[108] In July, the Select Committee on Energy Independence and Global Warming had invited George to respond to these concerns and discuss his company's efforts in more detail.[109]

The controversial debate that flared up in the context of this hearing illustrates how climate engineering gained political traction during the early 2000s by promising to realign action against climate change with economic benefits. Throughout his testimony, Russ George devised ocean fertilisation as a science-based tool that would turn environmental safeguarding into an 'enterprise':

> Our plan follows the consensus opinion of many ocean scientists, who have called for larger, more controlled, and more fully monitored iron addition

> trials that will generate the multidisciplinary data needed to understand this technology's true capacity as a tool for CO2 mitigation and ocean stewardship. As a for profit business, we are of course also interested in the economic implications of that data [...] We consider this work to be akin to the development of the applied science and technologies of agronomy and forestry and believe it can finally foster similar stewardship-based enterprises for the world's oceans.[110]

George devised Planktos' ocean fertilisation efforts as a 'frontier enterprise effort' that mobilised scientific research for the benefit of enterprise.[111] In this sense, he reasoned that 'if we succeed, we will have created a new industry. If we don't succeed, we will have produced a lot of great science'.[112]

Climate science, in other words, appears as a measure that can translate action against climate change into a business opportunity. More specifically, in George's observations, CO_2 becomes a form of negative commodity, a resource to be economically cultivated. Regarding the afforestation projects of Planktos' Hungarian subsidiary, KlimaFa, George suggested, for example:

> [...] if you're going to bank carbon dioxide in a carbon bank account, which is a forest, you need a bank guard. And we've selected the European National Park System as the bank guard for our carbon deposits that we're banking. We think that's safe, secure carbon.'[113]

By means of scientific expertise, George suggested that it would be possible to 'hire a tree or a green plant in the ocean to take that ton of carbon dioxide out of the atmosphere and turn it into those living plants, that living ecosystem [...], healing the harm done to date'.[114] Notions of techno-scientific control are directly linked here to environmental concerns about safeguarding nature. George continues: '[...] if we're lucky, if we do everything right, we might be able to mimic [the natural Galapagos bloom] and develop this as a technology that might have major utility in helping to reverse the decline of the ocean ecosystems'.[115]

Timothy Mitchell has shown how in the great engineering projects at the turn of the twentieth century, scientific expertise was politically envisioned as a means of 'taming' and 'ordering' nature.[116] We can see echoes of this vision for scientific expertise here, as climate engineering begins to appear as a project that would not only order, but essentially 'improve', or even 'heal', nature during the early 2000as. We will come back to this in more detail in the following chapter (see Chapter 6). Planktos' ocean seeding project became subject to fierce criticism and eventually got abandoned, as the company had issues securing the necessary funds. Nevertheless, the company's efforts illustrate how corporate interests joined the aforementioned federal interests in exploring ocean fertilisation as a measure to counteract climate change during these years and thereby importantly advanced a climate engineering-relevant expert infrastructure.[117]

This chapter has illustrated how climate engineering regained political currency between 2000 and 2009 when the very problem it promised to address was being reformulated. We have seen how climate change became problematised in the political realm as a challenge to techno-scientific innovation. The political problematisation of climate change did not concern the scientific prerequisites and epistemological underpinnings of a curious phenomenon anymore, but rather the effective management of an urgent problem. Building on the newly forged alliances between climate science and the state, which defined the 1970s through 1990s (see Chapter 4), climate science now further consolidated its status and evolved from a problem-defining to a problem-addressing authority during these years. Climate engineering thus moved to the surface of US climate policy during these years as it became part of an agenda to harness climate science (and particularly technology) as a tool for the state, as weaponry in the fight of a grand societal challenge.

NOTES

1 Hoffert in US House of Representatives, Subcommittee on Energy of the Committee on Science (2006: 91).
2 See, particularly, McCright and Dunlap (2011: 159); see also Brechin and Freeman (2004).
3 Fincham (2014).
4 Turner and Isenberg (2018: 168).

5 Keller (2009: 109).
6 See, particularly, McCright and Dunlap (2011: 159f.).
7 Keller (2009: 109).
8 See, e.g., McCright and Dunlap (2011).
9 Hoffert in US House of Representatives, Subcommittee on Energy of the Committee on Science (2006: 46); US House of Representatives, Committee on Government Reform (2006: 129). Hoffert had submitted this prepared testimony both to the House Science Committee, as well as the House Committee on Government Reform.
10 US House of Representatives, Committee on Government Reform (2006); see also: US House of Representatives, Subcommittee on Energy of the Committee on Science (2003: 47).
11 See, e.g., US House of Representatives, Subcommittee on Energy, Committee on Science (2003: 47); throughout discussion in US House of Representatives, Subcommittee on Energy of the Committee on Science (2006); throughout discussion in US House of Representatives, Committee on Government Reform (2006), US Senate, Committee on Environment and Public Works (2007: 66), US House of Representatives, Committee on Appropriations (2009: 239); US House of Representatives, Committee on Science and Technology (2010b: 32); US Senate, Committee on Energy and Natural Resources (2011: 36).
12 See, e.g., throughout discussion in US House of Representatives, Subcommittee on Energy of the Committee on Science (2006); throughout discussion in US House of Representatives, Committee on Government Reform (2006); US House of Representatives, Committee on Science and Technology (2009: 155); US House of Representatives, Subcommittee on Commerce, Justice, Science, and Related Agencies of the Committee on Appropriations (2009: 239).
13 US House of Representatives, Subcommittee on Energy of the Committee on Science (2006: 51).
14 Hoffert in US House of Representatives, Committee on Government Reform (2006: 124).
15 The 'first appearance' of these policy targets concerns the here examined corpus of documents, dealing with climate engineering measures.
16 US House of Representatives, Subcommittee on Energy of the Committee on Science (2006: 58).
17 Socolow in US House of Representatives, Committee on Government Reform (2006: 149).
18 United States of America (2007b).
19 Hoffert in US House of Representatives, Subcommittee on Energy of the Committee on Science (2006: 91).

20 Morseletto, Biermann, and Pattberg (2016: 3).
21 See, particularly Heintz (2012).
22 'Language, images, and numbers are different means of world creation. They aim at making visible what is not accessible directly, and therein attribute factuality to the displayed. This applies particularly to numerical representations [...]'. Heintz (2012: 7, translation J.S.). See also Luhmann (1990: 75).
23 Porter (2006: 1288).
24 Hulme (2014: 40, 43). See also: Morseletto, Biermann, and Pattberg (2016: 3); Weingart, Engels, and Pansegrau (2008: 13).
25 See, e.g., US Senate, Subcommittee on Oceans, Atmosphere, Fisheries, and Coast Guard of the Committee on Commerce, Science, and Transportation (2007: 71, 85).
26 US House of Representatives, Committee on Government Reform (2006: 73).
27 Fleming (2010: 253). According to a NOAA blog on climate change (Climate. gov), this sudden awareness might be reflective 'of post-9/11 fears about how fast the future could turn grim' (Fincham (2014)).
28 Previously, climate engineering had appeared on the US political record either by expert witnesses, or in some selected cases, by the executive branch (e.g., when in 2003, the Department of Energy gave notice on intent to prepare a so-called Programmatic Environmental Impact Statement (PEIS) for its carbon sequestration program).
29 Tom Davis in US House of Representatives, Committee on Government Reform (2006: 2).
30 For debates regarding the CCTP, see, e.g., US House of Representatives, Subcommittee on Energy of the Committee on Science (2003); US House of Representatives, Subcommittee on Energy of the Committee on Science (2006); US House of Representatives, Committee on Government Reform (2006). For discussion regarding carbon sequestration programs, see, e.g., US Department of Energy (2004); US Senate, Committee on Energy and Natural Resources (2007). Notably, we observe a conversation in the following, which was led by a Republican-dominated Congress, assessing the climate research program of a Republican administration that had proven notoriously averse to policy action on climate change. For the period between 2000 and 2006, when Tom Davis chaired the Committee on Government Reform, the committee was widely criticised for its lack of oversight over the Bush administration.
31 Barnett (2006).
32 US National Research Council (2012: 3).
33 CCTP's strategic plan promised that '[u]nder this new structure, climate change science and climate-related technology research programs are integrated to an extent not seen previously. The Climate Change Science Program (CCSP), led by the Department of Commerce, was established to reduce the uncertainties in climate

science and develop science-based resources to support decision makers. The Climate Change Technology Program (CCTP), [...] led by the Department of Energy, was formed to coordinate the Federal Government's portfolio of climate-related technology research and development activities, including technology deployment and adoption activities [...] and to focus efforts on the subset of priority activities that are part of the President's National Climate Change Technology Initiative' (US Climate Change Technology Program (2006: IV)).

34 See, e.g., Turner and Isenberg (2018: 168).
35 Barnett (2006).
36 Bodman qtd in Barnett (2006).
37 US House of Representatives, Committee on Government Reform (2006: 37); (US Climate Change Technology Program (2006: 122f., 127f.).
38 Chairman Davis in US House of Representatives, Committee on Government Reform (2006: 4, emphasis added).
39 US House of Representatives, Committee on Government Reform (2006: 2).
40 US House of Representatives, Committee on Government Reform (2006: 2).
41 US House of Representatives, Committee on Government Reform (2006: 2).
42 Van Atta in US House of Representatives, Committee on Government Reform (2006: 99–100).
43 US House of Representatives, Committee on Government Reform (2006: 2).
44 National Academies of Sciences (2007: 244).
45 Advanced Research Projects Agency (2017).
46 See, e.g., Czapla (2020).
47 Lane in US House of Representatives, Committee on Government Reform (2006: 84).
48 Lane in US House of Representatives, Committee on Government Reform (2006: 84).
49 Van Atta in US House of Representatives, Committee on Government Reform (2006: 119).
50 George in US House of Representatives, Select Committee on Energy Independence and Global Warming (2007: 93f.); Green in US House of Representatives, Committee on Energy Independence and Global Warming (2009: 40); US Senate, Subcommittee on Oceans, Atmosphere, Fisheries, and Coast Guard of the Committee on Commerce, Science, and Transportation (2007: 20, 24).
51 See, e.g., Schnare in US Senate, Committee on Environment and Public Works (2007: 64f.).
52 Andrews in US Senate, Subcommittee on Housing, Transportation, and Community Development of the Committee on Banking, Housing and Urban Affairs (2009: 9, 38).

53 Romm in US House of Representatives, Select Committee on Energy Independence and Global Warming (2007: 29, 36).
54 Romm in US House of Representatives, Select Committee on Energy Independence and Global Warming (2007: 28).
55 Figdor in US House of Representatives, Subcommittee on Energy and Environment of the Committee on Energy and Commerce (2009: 113).
56 SOLAS qtd. by Romm in US House of Representatives, Select Committee on Energy Independence and Global Warming (2007: 151); US House of Representatives, Subcommittee on Energy and Environment of the Committee on Energy and Commerce (2009: 113).
57 US Senate, Committee on Environment and Public Works (2007).
58 See, e.g., Office of Senator Whitehouse (2016).
59 Schnare in US Senate, Committee on Environment and Public Works (2007: 64ff.).
60 Schnare in US Senate, Committee on Environment and Public Works (2007: 72); see also Carlin qtd. by Schnare in US Senate, Committee on Environment and Public Works (2007: 123).
61 Schnare in US Senate, Committee on Environment and Public Works (2007: 84).
62 Barrett (2008); Carlin (2007).
63 See, e.g., Schnare in US Senate, Committee on Environment and Public Works (2007: 66).
64 Green in US House of Representatives, Committee on Energy Independence and Global Warming (2009: 40ff.). In contrast to Schnare, both Lee Lane and Kenneth Green were invited again to testify. For the testimonies of Lee Lane, see US House of Representatives, Committee on Government Reform (2006). For the testimonies of Kenneth Green, see US Senate, Subcommittee on International Development and Foreign Assistance, Economic Affairs, and International Environmental Protection of the Committee on Foreign Relations (2009); US House of Representatives, Committee on Energy Independence and Global Warming (2009); US House of Representatives, Select Committee on Energy Independence and Global Warming (2010); US House of Representatives, Subcommittee on Africa and Global Health of the Committee on Foreign Affairs (2010).
65 The hearings respectively address climate change adaptation in Africa in the USA and in 'particularly vulnerable nations' (US House of Representatives, Committee on Energy Independence and Global Warming (2009: 40ff.); US Senate, Subcommittee on International Development and Foreign Assistance, Economic Affairs, and International Environmental Protection of the Committee on Foreign Relations (2009: 26); US House of Representatives, Subcommittee on Africa and Global Health of the Committee on Foreign Affairs (2010: 60); US House of Representatives, Select Committee on Energy Independence and Global Warming (2010: 69f.)).

66 For George's account, see US House of Representatives, Select Committee on Energy Independence and Global Warming (2007); for the accounts of Conover, Doney, and Feely, see US Senate, Subcommittee on Oceans, Atmosphere, Fisheries, and Coast Guard of the Committee on Commerce, Science, and Transportation (2007).
67 Holdren merely suggested that 'all of the possibilities' of counteracting climate change need to be looked at, and 'that geoengineering ideas should continue to be studied and evaluated in the scientific community, in case something more promising can be identified' (Holdren in US Senate, Committee on Commerce, Science and Transportation (2009: 45)).
68 Figdor in US House of Representatives, Subcommittee on Energy and Environment of the Committee on Energy and Commerce (2009: 113).
69 Keller (2009: 8).
70 Sen. Mikulski in US Senate, Committee on Environment and Public Works (2007: 77).
71 Schnare in US Senate, Committee on Environment and Public Works (2007: 151f.).
72 Keller (2009: 95).
73 Keller (2009: 96).
74 Maasen and Weingart (2006: 6).
75 Keller (2009).
76 For the rise of the notion of scientists as managers or pilots, see also Bonneuil and Fressoz (2016: 87f.).
77 US House of Representatives, Committee on Government Reform (2006); US Senate, Committee on Environment and Public Works (2007).
78 See, e.g., Schnare in US Senate, Committee on Environment and Public Works (2007: 66); Lane in US House of Representatives, Committee on Government Reform (2006: 84–85).
79 NAS qtd. by Lane in US House of Representatives, Committee on Government Reform (2006: 85); US National Academy of Sciences (1992: 433f.).
80 Barrett in US Senate, Committee on Environment and Public Works (2007: 109–20).
81 Carlin in US Senate, Committee on Environment and Public Works (2007: 121–27).
82 Carlin in US Senate, Committee on Environment and Public Works (2007: 122).
83 Schnare in US Senate, Committee on Environment and Public Works (2007: 64, 65, 67). For comparative economic perspectives in later years, see, e.g., US House of Representatives, Committee on Science and Technology (2009: 34f).
84 Referring to the Buesseler et al. study, Senator Culbersome suggested R&D into ocean fertilisation measures to prevent US politics from 'driv[ing] America

back to the industrial production levels of 1920' (US House of Representatives, Subcommittee on Commerce, Justice, Science, and Related Agencies of the Committee on Appropriations (2009: 305)).

85 Schnare in US Senate, Committee on Environment and Public Works (2007: 71).
86 Barrett in US Senate, Committee on Environment and Public Works (2007: 113).
87 US House of Representatives, Select Committee on Energy Independence and Global Warming (2007); US Senate, Committee on Energy and Natural Resources (2007); US Senate, Subcommittee on Energy and Water Development of the Committee on Appropriations (2009).
88 US Department of Justice, Antitrust Division (2003: 16552). For reference of the project, see also, e.g., US House of Representatives, Committee on Government Reform (2006: 95). In 2011 and 2013, DuPont and Bank of America would join the partnership (US Global Climate and Energy Project 2017b).
89 Turner and Isenberg (2018: 33).
90 US Global Climate and Energy Project (2017a).
91 Antitrust Division, US Department of Justice (2003: 16552ff.).
92 US House of Representatives, Subcommittee on Energy of the Committee on Science (2003: 24); US Senate, Committee on Energy and Natural Resources (2007: 20, 28); US House of Representatives, Select Committee on Energy Independence and Global Warming (2007: 95); US Senate, Committee on Environment and Public Works (2007: 69).
93 US Senate, Committee on Energy and Natural Resources (2007: 45).
94 See, US House of Representatives, Subcommittee on Energy of the Committee on Science (2003: 21); US Senate, Committee on Energy and Natural Resources (2007: 20).
95 See, e.g., US Senate, Committee on Energy and Natural Resources (2007: 20).
96 See, e.g., US House of Representatives, Subcommittee on Energy of the Committee on Science (2003: 6, 21); US Senate, Committee on Energy and Natural Resources (2007: 28).
97 See, e.g., Chadwick and others (2006); Equinor (2019); Massachussetts Institute of Technology (2019).
98 US Department of Energy (2017).
99 US Department of Energy (2004: 21515).
100 US Department of Energy (2004: 21515).
101 US Department of Energy (2017). In this context, the DOE also supported research into carbon capture and storage at the Sleipner gas field (US Department of Energy, Office of Fossil Energy (2009)).
102 US Department of Energy (2004: 21514).
103 US Department of Energy (2004: 21515).

ASSEMBLING AN ENGINEERING PROBLEM

104 US Department of Energy (2004: 21516). The program defines Oceanic Sequestration as 'a range of technologies and methods employed to bind, store, or increase carbon dioxide uptake in the ocean. Such technologies may include deep ocean injection of captured carbon dioxide gas or the enhancement of free carbon dioxide uptake by marine ecosystems through ocean fertilization or other methods to enhance natural absorption processes' (US Department of Energy (2004: 21514)).
105 International Maritime Organization (2007).
106 International Maritime Organization (2007); see also: Sagarin and others (2007: 7); Abate and Greenlee (2009: 558).
107 Russ George strongly disagreed with this narrative in his congressional testimony some weeks later: 'what we have is we've received one fax from the EPA asking me to phone them. I telephoned them. I had about an hour long, informal phone conversation with them. The very next thing I heard [...] about the EPA, was I received a telephone call from a reporter in Ottawa, Canada who said that a very radical environmental group called ETC based in Ottawa had handed the reporter the contents of the EPA's presentation to the London Dumping Convention meeting in Spain that was taking place at that very moment and would I comment on the EPA's criticism of our work'. (George in US House of Representatives, Select Committee on Energy Independence and Global Warming (2007: 150)).
108 International Maritime Organization (2007). The United States Ocean Dumping Act is the statute implementing the London Convention.
109 US House of Representatives, Select Committee on Energy Independence and Global Warming (2007: 150, 92); US Senate, Committee on Environment and Public Works (2007: 69).
110 George in US House of Representatives, Select Committee on Energy Independence and Global Warming (2007: 93).
111 US House of Representatives, Select Committee on Energy Independence and Global Warming (2007: 95).
112 George in US House of Representatives, Select Committee on Energy Independence and Global Warming (2007: 143).
113 US House of Representatives, Select Committee on Energy Independence and Global Warming (2007: 153).
114 US House of Representatives, Select Committee on Energy Independence and Global Warming (2007: 151).
115 George in US House of Representatives, Select Committee on Energy Independence and Global Warming (2007: 147).
116 See, particularly, Mitchell (2002).
117 See also Abate and Greenlee (2009: 558).

6

DEVISING A PROJECT OF CLIMATOLOGICAL CULTIVATION AND CONTROL

WE HAVE NOW COME FULL CIRCLE TO WHERE WE STARTED AT THE OUTSET of this book: as we approach the second decade of the new millennium, climate engineering incrementally cemented its presence in US climate policy. By the end of 2009, Congress embarked on its formal inquiry into the issue, triggering the first peak in political attention to these controversial measures (see Fig. 2.1). In contrast to the early 2000s, this meant that climate engineering became established as an issue in its own right within the political realm. It continued its trajectory once again by shifting its status, evolving from a controversial techno-fix to a basic component of a national climate science policy agenda.

Part I of this book set the stage of our analysis in this context. It depicted the conflicted status of climate engineering during this important historical moment and thereby raised the guiding puzzle of this book. We saw, how, despite being framed as 'a bad idea'[1], climate engineering became programmatically assessed and was internalised into the federal infrastructure as a potential remedy against anthropogenic climate change during these years.

Building on the previous chapters, we can now return to this analytical starting point and address the puzzle raised here. Instead of explaining the controversial arrival of climate engineering on political agendas in 2009 with sheer urgency in the face of the daunting climate catastrophe – as a last resort or Plan B – we can make sense of how we got here by turning to the historically contingent science-state alliances, defining yet another chapter in the chequered career of these measures. With this frame of analysis, the apparently conflicted status of climate engineering can be understood as a kind of *synthesis* – one that reconciles two historically conflicting roles of climate science within the state. Framed as a 'bad idea whose time has come', climate engineering caters to visions of techno-scientific control over the climate all the while anticipating a critique of such measures. It builds on the promise of science as a tool for the state, all the while connecting to 'green' notions of environmental safeguarding. The contested status of climate engineering emerging in 2009 in this sense aligns the hopes for control that had shaped political interest in climate modification for the first half of the twentieth century, with the positions of climate scientists and environmental movements that questioned precisely these hopes for control during the second half of the twentieth century.

In this chapter, we will revisit and expand on the analysis provided in part I of the book. We will zoom into the science-state alliances that defined the career of climate engineering from the years leading up to the official inquiry into climate engineering in 2009 to around the end of the Obama administration in 2016. In doing so, we will contextualise the last resort narrative and see how science and politics came together during these years by formulating climate engineering as a *project of climatological cultivation and control*.

As in the previous chapter, we will begin by taking a brief glance at the political landscape of the time. We will turn to the incoming Obama administration as the defining political environment in which the career of climate engineering took shape during these years. For the remainder of the chapter, we then shift our gaze from the political environment to the experts and expertise that defined this stage in the career of climate engineering. We will explore the role of scientific expertise in devising this option of deliberately intervening in and controlling the Earth's climate. We will see how natural and social scientific modes of observation have essentially assembled this project of climatological cultivation and control. The chapter then maps the corresponding expert infrastructure undergirding these expert modes of observation and introduces the experts and expert organisations that formulated this project of climatological cultivation and control.

CLIMATE ENGINEERING BECOMES PART OF A NATIONAL CLIMATE SCIENCE POLICY AGENDA

The formal US political inquiry into climate engineering, beginning in November 2009, coincided with rather substantial political shifts in the United States. After eight years of a Republican government that was 'officially sceptical'[2] of climate change and opposed to regulating carbon emissions, the incoming Obama administration promised a change of direction. Even before the new administration took over the Oval Office in January 2009, policy change seemed imminent. Congressional hearings on environmental issues and climate change 'virtually exploded' two years earlier during the so-called Democratic wave of 2007, when Democratic majorities returned to both the Senate and the House of Representatives for the first time since 1995.[3] And when John McCain and Barack Obama, two Senators who agreed on the urgency of the issue, were competing in the 2008 presidential campaign, bipartisan consensus and policy action on climate change appeared to many commentators at the time as 'almost inevitable'.[4] Domestic and international expectations for policy change and real action on climate change were high and many anticipated a potentially leading role for the United States in a newly concerted international effort to tackle this urgent issue.[5]

With the benefit of hindsight, however, the climate policy legacy of the Obama administration appears rather mixed.[6] In the years that followed his inauguration, climate change regressed into the quintessential partisan issue – a development which would eventually reach its peak a couple of years later when Donald Trump was elected the 46th president of the United States.[7] In their analysis of the Obama administration's climate policy agenda, Graciela Kincaid and Timmons Roberts demonstrate that the topic of climate change lost political traction during Obama's years in office and was instead replaced by references to energy issues or the environment in general.[8] One reason for this was a somewhat forced fight for bipartisanship. As the Democrats lost control over Congress after Obama's first two years as president, the administration had to foster bipartisanship on critical policy issues. In this fight for bipartisanship, the administration felt that climate change 'needed some time off because it had gotten so tainted and polluted".[9] This increasing partisan contestation proved persistent and ongoing. By the time of the 2012 presidential debates, which would prepare Obama's second term, there was no mention of the climate change issue at all for the first time since 1984.[10] Fast forward another four years, and this dynamic further intensified. With the election of Donald Trump into the White House in 2016, the ambiguous status of the climate change issue escalated into outright denial of its existence. What followed since has been widely characterised as a full-fledged assault on climate science and even basic recognition of the urgency of this problem.

Against the backdrop of this dynamic political environment, we can now revisit and historically situate the emerging politics and contested status of climate engineering around 2009. Specifically, we can add another layer to the picture drawn in Part I of the analysis by turning to the status of climate science in politics that defined this particular historical setting in the career of climate engineering.

In contrast to the height of the Bush years, the political exploration of climate engineering followed a less blatantly techno-optimistic tone. On the surface, this produced a somewhat contradictory effect: the formal inquiry into climate engineering came with an openly critical assessment of its merit. Policymakers as well as invited expert witnesses appeared outright sceptical of climate engineering, as we saw in Part I of the analysis. Our glance at the political environment in which climate engineering took shape during these years suggests that this negative assessment of the proposed measures is not necessarily contradictory but might be explained as a response to a Democratic constituency that expected real action on climate change – that is, the mitigation of anthropogenic emissions and a respective change in behaviour.

Following this line of reasoning, climate engineering was rejected as a political project that might delay action on climate change. It was no longer explored as a controversial technological apparatus to control the Earth's climate as technological weaponry at the hands of the state (see Chapter 5). Instead, climate engineering now became formulated as 'just science'. Beginning already with the Democratic wave of 2007, climate engineering began to move from highly contested and politicised debates that we explored in the previous chapter to rather technical scientific debates. It became increasingly normalised as a basic component of a national climate science agenda – an agenda that would not only help to decipher, but also govern and cultivate the climate. An agenda, in other words, that saw climate science as a relevant governance tool.

In the years leading up to the 2009 official inquiry into climate engineering, this shift became particularly evident in the political exploration of ocean fertilisation measures (see also Chapter 5).[11] Climate engineering was formulated here as part of a science policy agenda that sought to make climate science more relevant to the needs of political decision makers. Cast in this light, climate engineering hardly seemed to propose a politically controversial or radically new approach, but rather appeared as the next logical step in developing a comprehensive national climate science agenda.

In May 2008, for example, the Democratic-controlled Senate suggested a comprehensive assessment of ocean fertilisation measures as part of a proposed bill that was meant to update the national global climate change program of the United States. Although it was never enacted, the so-called Global Change Research Improvement Act of 2007 sought to improve the US Global Change Research Program (USGCRP) from 1990. This was a program that for the first time had set out to coordinate climate change research across federal agencies in the United States, as we saw in Chapter 4. The goal of this improvement was to foster federal research capacities that would 'produce information' that could 'better meet [...] the expressed needs of decision-makers'.[12]

Climate engineering became formulated in this context as part of a climate science agenda that sought to develop climate science as a central 'information-base' or 'decision-making tool' for political decision-makers. These political decision-makers were addressed here as 'resource managers [that] require accurate, relevant, timely, and user-friendly data on climate change [...]'.[13] Climate science, in other words, was envisioned as critical governance knowledge; it appeared as providing a perspective that would 'assist the Nation and the world to better understand, assess, predict, mitigate, and adapt to the effects of human-induced and natural processes of global change'.[14] The scientific deciphering of climate change promised its political manageability.

DEVISING A PROJECT OF CLIMATOLOGICAL CULTIVATION AND CONTROL

At the heart of this document thus laid the call for a better scientific understanding of climate change as a distinctly national-political concern. And this included not only higher resolved observations, improved measuring, and monitoring of climate change, but also an assessment of 'existing research, potential risks [...], and the effectiveness' of fertilising the oceans to counteract climate change.[15] Chapters 1 and 2 furthermore suggested how between 2009 and 2011, policymakers began internalising climate engineering into the federal climate science infrastructure. The National Oceanic and Atmospheric Administration (NOAA) and other scientific and independent agencies were being tasked with advancing climate engineering research, effectively integrating these approaches into the national climate science agenda. We have seen how climate engineering appeared here in the form of basic scientific challenges and became a direct policy concern.

To sum up, around 2009, climate engineering became part of a climate science agenda that envisioned climate change as a project of cultivation by human rationality, ingenuity, and reason.[16] As we will explore in more detail in the following sections, these measures became established and institutionalised as a means to decipher, cultivate, and control the climate.

WHEN WE BRIGHTEN THE CLOUDS, WE SEE THAT THE PLANET COOLS: EXPERT MODES OF OBSERVATION

> Wind-driven spray vessels will sail back and forth perpendicular to the local prevailing wind and release micron sized drops of seawater into the turbulent boundary layer beneath marine stratocumulus clouds. The combination of wind and vessel movements will treat a large area of sky. When residues left after drop evaporation reach cloud level they will provide many new cloud condensation nuclei giving more but smaller drops and so will increase the cloud albedo to reflect solar energy back out to space.
>
> Salter et al. 2007, qtd. by Lee Lane, testifying before the House Science Committee[17]

In the following, we turn to the kinds of expertise that made the project of climatological cultivation and control politically 'legible'.[18] We will unpack the expert modes of observation which the political record on climate engineering, the congressional debate and political internalisation of this issue, rested on.[19] In doing so, we will see how two such modes of observation established this project of engineering the climate in the political sphere – namely, *climate models* and so-called *natural analogies*. These two modes of observation effectively assembled the abstract notion of climate engineering into a potential technology; they realised climate engineering as a set of interventions that are technically working. And they generated strikingly specific and concrete observations regarding the effects of deploying such interventions.

Climate models, to begin with, have not only been paramount to scientific endeavours to understand and predict climate change, but they also provided the essential grounds for the political problematisation of this issue for at least half a century.[20] The policy documents examined in this book suggest how the expansion of modelling efforts have become a national strategic matter, answering to 'emerging national needs' such as the building up of national resilience and adaptation capacities against potentially catastrophic climate change.[21] To use a concept by Bentley Allan, climate models have provided an essential device to 'translate' the issue of climate change 'into a portable, global object' which could be communicated and problematised in various societal contexts.[22] Climate models in this sense serve to abstract; they transform the tremendously complex and multi-layered phenomenon of a changing climate into an observable process. They provide formalised transcriptions which isolate distinct properties of this multi-facetted phenomenon.[23] In this capacity, climate models have proved critical in furnishing observations of a changing climate with hopes of its deliberate intervention and control. In such models, a changing climate not only becomes observable, but it appears as a set of deciphered causal connections, accessible to deliberate manipulation and control.

Climate models have therefore come to provide the core epistemological ground for examining climate engineering – and especially solar radiation management – as a potential approach to counteracting climate change. They appear as an essential political decision-support tool in this context.[24] Prominent examples such as the Geoengineering Model Intercomparison Project (GeoMIP) or the Carbon Dioxide Removal Model Intercomparison Project (CDRMIP) demonstrate how, over the years, rather isolated modelling studies grew into increasingly complex and internationally coordinated simulations including multiple models.[25]

Alongside models, so-called natural analogues have provided an essential expert mode of observing and assessing climate engineering. In contrast to climate models, these natural analogues are expert observations that compare climate engineering measures to 'naturally occurring' phenomena.[26] In the context of solar geoengineering, these analogues are typically volcanic eruptions – Mt. Tambora[27], Mt. Kasatochi[28], Mt. El Chicón[29], and most prominently, the eruption of Mt. Pinatubo. Over the years, the eruption of Mt. Pinatubo has reached symbolic status in terms of suggesting how solar radiation management would work.[30] Carbon dioxide removal measures, on the other hand, have been compared to trees and other 'natural' processes of carbon sequestration.[31]

By zooming into some of the expert testimonies on climate engineering, we will unpack how both modes of observations – climate models and natural analogues – essentially served to assemble an abstract techno-scientific concept into a concrete piece of applicable technology. This means that even though most of the discussed climate engineering approaches may not have existed as pieces of material hardware at the time of these assessments, these modes of observation have served as a kind of expertise that effectively envisioned these technologies within the political process, making them accessible to political decision-making. These modes of observation, in other words, served to create a vision of the 'global thermostat' that climate engineering promises to provide.[32]

Establishing a technology that works

Climate models and natural analogues first served to establish causal closure in the congressional exploration of climate engineering. This means these modes of observation have served as empirical evidence that climate engineering – and especially solar radiation management – technically works:

> [...] both observations of the response of climate to large explosive volcanic eruptions [...] and all modelling studies conducted so far [...] show that

with sufficient stratospheric sulfate aerosol loading, backscattered insolation will cool Earth.[33]

Climate models and natural analogues thus effectively translate the complex issue of anthropogenic climate change into distinct climatological mechanisms. And by isolating these mechanisms, the issue of climate change appears as amenable to targeted intervention and control. The multifaceted societal roots of the problem, the economic structures and industrial processes, as well as its diverse consequences, are all boiled down to a set of climatological mechanisms.

Natural analogues have primarily served as a 'natural' 'proof of concept' in this context.[34] They have served to translate the otherwise abstract notion of climate engineering into something that already happens 'naturally'. In these observations, solar radiation management, for example, becomes '[...] the identical process that happens when volcanoes erupt and they cause cooling'.[35] By translating the abstract into such 'naturally occurring' phenomena, natural analogues establish climate engineering as both harmless and plausibly effective. In this mode of observation, climate engineering emerges as 'working' *for everyone to see*.[36] One of the expert witnesses, for example, explained to Congress that

> There are questions about how good a short-term eruption is as an analogue for a continuous injection of material into the stratosphere. Nevertheless, the natural experiment of volcanic eruptions gives us confidence that the approach will basically work, and while there might be negative consequences, the world will not come instantly to an end, and that after stopping a short-term deployment, the world is likely to return to its previous trajectory within years.[37]

Drawing from analogous observations, experts suggested that 'we know this [climate engineering] basically works'.[38] The epistemic authority of analogous observations was presented as greater and more conclusive than the effectiveness of mitigation efforts:

DEVISING A PROJECT OF CLIMATOLOGICAL CULTIVATION AND CONTROL

> [...] because these techniques mimic natural phenomena, *we know more about how quickly and well they work than we do about the efficacy of attempting to reduce greenhouse gases.* We have measured the effects of the natural processes and can state with considerable certainty, bordering on complete certainty, that they will produce the result sought.[39]

Natural analogues furthermore serve to suggest the possibility of enhancing nature in this context. Climate engineering appears not only as mimicking 'natural' cause and effect mechanisms; it is devised as optimising these mechanisms, it appears as 'optimised nature':

> The sulfur-containing particles thrown out by eruptions are probably less than optimal. It appears reasonable to believe, however, that humans could improve on nature substantially by refining the type of particles used and minimizing other possible environmental side effects with a little research and development.[40]

This suggestion to 'enhance' nature was also applied to carbon removal approaches:

> Left to its own devices, nature will take on the order of a hundred thousand years to reabsorb and fixate the excess carbon that human activities have mobilized and injected into the atmosphere. The purpose of mineral sequestration in managing anthropogenic carbon is to accelerate these natural processes to the point that they can keep up with human carbon dioxide releases.[41]

Modelling observations complement these 'natural' proofs of concept by suggesting the possibility of targeted application. Simulating a targeted climate intervention implies not only being able to identify and isolate the relevant physical mechanisms for climatic change, but it implies being able to purposefully reproduce these mechanisms. Climate models, in this sense, are essential to this project of climatological intervention and control. They make the concept of climate engineering into a 'technology' in the sense that they not only suggest understanding, but also the possibility of control.[42]

Experimenting with an engineered climate

Secondly, climate models and natural analogues serve to provide seemingly controlled empirical observations regarding the hypothetical deployment of climate engineering. While natural analogies serve as 'natural experiments', climate models are mobilised as theoretical ones.[43] Both modes of observation allow trials without actually deploying the technology; they provide a virtual testing ground for a technological concept that remains 'too dangerous, too expensive, or perhaps impossible to perform with the real thing', to use Schneider's words.[44]

In fact, following these expert observations, models and volcanoes become 'the real thing'. There is no meaningful epistemological difference in the way that these observations are mobilised: '[I]n climate models when we brighten the clouds, *we see* that the planet cools. When we inject an aerosol like volcanoes do, *we see* that the planet cools'.[45]

Climate models and natural analogues thus make the hypothetical application of climate engineering accessible to strikingly concrete and specific empirical observations. In these observations, climate engineering becomes evident not only in its basic climatological mechanisms, but even in its future consequences. Drawing from these modes of observation, experts suggest specific insights into the consequences, side effects, and potential risks of a future climate engineering deployment. The risk of draught or acid rain[46] becomes as observable as a sudden cessation of a SRM scheme[47], its potential impacts on the Indian monsoon, on droughts in Africa, or on the oceanic biosphere:[48]

> There are also undesirable things that happen. We see that even though we might make the average temperature of the planet about right, the rainfall patterns would change some from today, and some places become warmer and some places become cooler.[49]

DEVISING A PROJECT OF CLIMATOLOGICAL CULTIVATION AND CONTROL

Climate models in this context allow observations of a hypothetical climate engineering deployment. Models, for example, seem to make the experimental replication of a Mt. Pinatubo eruption every other year precisely observable:[50]

> While the aerosols are located above the poles, they would shield the sea ice to keep the poles cooler in summer, and then allow the aerosols to disappear during winter when there is no sunlight at the poles anyway. Robock (2009) has shown that the particles *actually* spread and produce a cooling beyond the Polar Regions.[51]

Following this line of reasoning, one expert implied that his experiments suggested

> [...] that clouds injected into the Arctic stratosphere would be blown by winds into the mid-latitudes and would affect the Asian summer monsoon. Observations from all the large high latitude volcanic eruptions of the past 1500 years, Eldgjá in 939, Laid in 1783, and Katmai in 1912, support those results.[52]

These observations very much illustrate the 'uneasy epistemic space' of models, as described already in the 1990s by Sergio Sismondo.[53] Models cross the boundaries of measuring device and data, of theory and experiment. Models, in this sense, work as analogues. They are '[...] studied in the way that natural systems might be:'[54] their insights are mobilised as quasi-empirical.

A note on interdisciplinarity

This particular formulation of climate engineering as a project of climatological cultivation and control is thus not only built on detailed and precise climatological observations, but importantly, it also perpetuated the need for them. As the politics of climate engineering directly translated into basic challenges in the atmospheric sciences, the political capacity to decide and act on climate change seemed to directly rest on better modelling and measuring capacities.[55]

Such modelling and measuring capacities were not only seen as essential for exploring technological viability,[56] but also for crafting sensible governance and providing monitoring and management capacities.[57] The emerging politics of climate engineering in this sense entailed addressing 'the holy grail' of climate science – that is, 'using present observations to predict future climate states' or 'the science of fingerprinting'.[58]

This prominence of climatological expertise and the respective modes of observations, however, should not imply a lack of social scientific expertise in these politics of climate engineering. In fact, assembling this project of climatological cultivation and control essentially rested on social scientific and ethical expertise.

This is maybe most clearly displayed in the prominent role of economic expertise throughout the career of climate engineering. Economic expertise has played a critical role in assessing the potential of climate engineering as a policy tool throughout the years, as we have seen in previous chapters. It has provided one of the most prominent sources of expert advice and is a critical mode of evaluating, comparing, judging, and deciding on the merit of these measures as a potential policy approach to tackling climate change. Reference to 'cost' has served as a critical mode of both promoting and contesting climate engineering measures in this context (see also Chapters 1, 3 and 5).[59] Experts argued, for example, that 'the only reason that we are considering doing geoengineering [...] is because the consequences of not doing anything might be more costly'.[60] Or they explained, 'it makes sense for us as American taxpayers to invest some of our hard-earned dollars in exploring ways to cost-effectively reduce the environmental threats that are facing us'.[61] Such observations effectively devised the issue of climate change as a challenge of 'cost-effective risk management'.[62] With the official inquiry into climate engineering, economic observations began providing a central mode of elaborating on and differentiating the distinct challenges that climate engineering would entail.

In Chapter 1 we saw that experts deemed cost as particularly relevant in assessing the political viability of carbon dioxide removal methods, even arguing that 'cost is likely to be the primary consideration governing deployment' of these approaches.⁶³ But also for solar radiation management approaches, uncertainty in future costs appeared as a core challenge. 'Cost' in this context concerned monitoring expenses and liability issues, but also 'social costs' such as risks or 'the cost of public opinion'.⁶⁴ Game-theoretical perspectives that have grown increasingly prominent within climate policy beginning in the 1990s, served to compare conventional mitigation measures with the techno-scientific 'making' of climate in this case of solar radiation management. Risk of free-riding or suboptimal coordination of behaviour, on the one hand, was confronted with substantial and unpredictable technological risks as well as the danger of unilateral action with global consequences, on the other. The figure of a 'rogue state' also played a critical role in this context.

Apart from these economic observations, experts and policymakers continuously emphasised the need for regulatory, ethical, and political or social scientific expertise more generally: 'it is important to acknowledge that climate engineering carries with it [...] ethical and political concerns'⁶⁵ or that 'the legal, governmental, socio-political and ethical issues may ultimately be greater challenges to deployment'.⁶⁶ Yet, this continuous emphasis on the need for interdisciplinarity hardly meant that policymakers invited a diversity of critical, competing, or challenging perspectives. Instead, social scientific, economic, and ethical observations largely *followed* climatological observations (and not the other way around). In other words, this social scientific expertise served as accompanying research – as essential expertise to advance and realise this project of climatological cultivation and control.⁶⁷

In the accounts of the experts, such social scientific expertise would be essential, for example, for deciding 'whose hand would be on the thermostat'⁶⁸ or 'what temperature [...] we want the planet to be'. For example, 'do we want it to stay constant? Do we want it to be at 1980 levels, do we want it at 1880 levels'? and 'who decides? What if Russia and Canada want it a little bit warmer and India wants it a little bit cooler'?⁶⁹ Such questions hardly confront, but rather reinforce and give shape to this notion of an engineered and technologically controlled climate. They confine the scope of ethical, normative, and social scientific concerns to the question of governing the already assembled global thermostat.

This escorting or accompanying role of social scientific and ethical perspectives is also suggested in the common reference to 'public engagement strategies' as a meaningful remedy to tackling normative concerns regarding climate engineering.[70] One expert witness, for example, argued that 'we need to find a way to engage the opinions of a very diverse group of people on the planet so *that this can be done* in an orderly and acceptable manner'.[71] What 'this' is seems already clear and predetermined.

Social scientific modes of observation thus have not systematically challenged the very notion of targeted climate intervention, but rather served to elaborate and differentiate its pursuit.[72] By qualifying regulatory, governance, and legal challenges in the development and deployment of targeted climate intervention, they, too, have proved essential in advancing this project of climatological cultivation and control.

To sum up, these observations substantiate what the first part of this book could only hint at, namely that science and politics is not only coupled via organisations and experts, but also via relevant expertise. These expert observations illustrate how the political 'career' of climate engineering is directly bound to processes of scientific knowledge production. The trajectory of this concept as a policy measure is coupled with observational and measuring devices; it is linked to scientific modes of observing and making sense of climate change. This outlook suggests that the consequences of climate engineering research hardly amount to environmental impacts. We should pay attention to these modes of observation for gaining a more differentiated understanding of how research shapes future technologies and defines likely trajectories. While it is obviously crucial to anticipate potentially harmful environmental side effects of climate engineering field studies, this should not curtail a more comprehensive understanding of the various dimensions in which climate engineering research matters societally.

UNPACKING THE EXPERT INFRASTRUCTURE

As we have seen throughout the previous chapters, climate engineering did not (re)gain political traction around 2009 simply because experts or policymakers placed it on the agenda. Instead, the emergence of this new governance object, the 'politicisation' of climate change as an issue of climatological cultivation and control, followed a complex translation process in which the political system internalised perspectives and expert modes of observations from its societal environment. Against this backdrop, I want to suggest that the political selection of expertise is a critical part of the above-mentioned translation process. To quote Reiner Grundmann, expertise is always delivered 'at the request of someone else'.[73] In this final section of the chapter, we will revisit the two arenas in which climate engineering materialised in US politics around 2009 and unpack how scientific expertise precisely connects to politics here. We will turn to the climate engineering expert infrastructure, i.e. the routes and channels that the political system established to 'request' expertise – that is, to internalise, consider, and quite literally, hear scientific experts.

Staged advice: Scientists as political spokespeople?

In Chapter 2, we saw how scientific expertise has come to shape the emerging politics of climate engineering *by being invited* to do so. The political system has established a number of different channels that effectively guide the selection and flow of external expertise into the political system in this context. Policymakers have invited experts to testify before Congress; they have requested experts to inform the legislative assessments and federal inventory and they have commissioned scientific assessments. Table 7.1 provides an overview of all experts that were invited to provide advice on climate engineering in the context of establishing an 'official record' on the issue between 2009 and 2017.

I have suggested that this invited expertise serves as a kind of 'staged advice' (see Chapter 2). Now we can further unpack the role and status of this 'staged advice' as a particular component of the climate engineering expert infrastructure and as a distinct setting in which scientific expertise shapes politics.

The notion of 'staged advice' seeks to emphasise the double-sided character of this particular form of advice. On the one hand, the congressional inquiry, as well as the scientific assessments, represent critical nodes of linking scientific observations to political observations. Table 7.1 suggests that both formats

ENGINEERING THE CLIMATE

Name	Background	Hearings 2009–2010	Hearing 2017	GAO 2010	GAO 2011	sum	NASEM (2015)	Royal Society (2009)	sum	TOTAL APPEARANCE
		INFORMING CONGRESSIONAL INQUIRY					INFORMING SCIENTIFIC ASSESSMENTS			
Caldeira Ken	science	x		x	x	3	x	x	2	**5**
Rasch Philip	national laboratory	x	x	x	x	4	x		1	**5**
Fleming James	science	x		x	x	3	x		1	**4**
Morgan Granger	science	x		x	x	3	x		1	**4**
Shepherd John	science	x		x	x	3		x	1	**4**
Barrett Scott	science	x		x	x	3			0	**3**
Keith David	science	x			x	2		x	1	**3**
Robock Alan	science	x		x	x	3			0	**3**
Doney Scott	national laboratory				x	1	x		1	**2**
Fetter Steve	federal bureaucracy				x	1	x		1	**2**
Lackner Klaus	science	x			x	2			0	**2**
Lee Lane	think tank	x			x	2			0	**2**
Long Jane	national laboratory	x			x	2			0	**2**
MacCracken Michael	think tank			x	x	2			0	**2**
Redgwell Catherine	science			x		1		x	1	**2**
Seidel Stephen	think tank			x	x	2			0	**2**
Victor David	science			x	x	2			0	**2**

TABLE 7.1 Staged Advice (2009–2017). Overview of experts who have been invited to inform the establishment of an 'official record' on climate engineering in US climate policy between 2009 and 2017 (see Table 3.1) in at least two different contexts. The experts are ranked according to the frequency of their appearance.

have provided critical mechanisms to channel expertise from academic research contexts to the political realm. On the other hand, scientific expertise connects to policy making in this arena of staged advice precisely by reinforcing a clear boundary between science and politics. Science studies scholarship has suggested that this, in fact, is a key dimension in the provision of scientific advice in general: trust in expert judgement is precisely generated by suggesting a clear divorce between the evidence base that scientific expertise provides, on the one hand, and the decision that policymakers take, on the other.[74] The advisory process then becomes an arena of boundary-work between science and politics.

The so-invited experts occupy a particularly prominent role in the emerging politics of climate engineering. Be it via congressional testimonies, commissioned reports, or scientific assessments, these experts are invited to co-define the stakes of the issue and set the terms of the debate. In particular, the congressional experts who are invited to testify and inform the legislative assessments appear as highly visible. They have shaped the emerging politics of climate engineering by quite literally 'speaking' to politics. In this speaking capacity, they essentially also shape the second arena (as suggested in Chapter 2) because by co-defining the stakes of the debate over climate engineering and determining the technicalities and feasibility of the issue at hand, these congressional experts essentially guide the political inventory and structure the political internalisation of climate engineering into the federal bureaucracy. And what is more, in their capacity to assemble climate engineering as an epistemic object, they also serve as a kind of gatekeeper for determining relevant expertise. By adding papers or position statements to the record or otherwise referencing expert observations, these congressional experts effectively decide which kinds of evidence are relevant for the issue at stake.

In the following, we will take a closer look at who gets to speak in this arena of staged advice. To do so, we will distinguish between the selection of experts via scientific assessments and via the congressional inquiry.

Selection of experts via scientific assessments

In the case of scientific assessment reports, science connects to politics through particular organisational procedures. Scientific assessment reports are usually the product of a highly formalised, almost ritualistic process which seeks to make sure that scientific insights can be harnessed for national needs all the while safeguarding the integrity of the scientific process.[75] Accordingly, the scientific assessments on climate engineering have been generated by a rather big pool of different expert voices. Table 7.1 suggests that there is only one scientist who informed both of the featured assessment reports. The remaining experts contributed to just one of the here considered publications. This picture generally attests to a rather impersonal selection procedure.

Gupta and Möller argue that these scientific assessments 'leverage and reflect the scientific eminence associated with the institutional context from which they emerge, which serves to endow them with epistemic authority and legitimacy […]'.[76] National scientific academies provide one particularly important institutional context in this regard. These academies essentially pool scientific expertise that is generated by a decentralised national scientific system of universities and research institutes. The mission of the US National Academy of Sciences, for example, has been

> to improve government decision making and public policy, increase public understanding, and promote the acquisition and dissemination of knowledge in matters involving science, engineering, technology, and health.[77]

Sponsors, such as the federal government, can request studies on subjects they wish to be informed on.[78] The Academy's board members then suggest pertinent experts for the compilation of the assessment reports. While the production process of these reports is somewhat similar to scientific publications, such as peer-reviewed papers or monographs, the goal of the respective study is usually clearer than would be the case in academic research or in the writing process. This predetermined goal and the distinct scope of the study are defined by both the report's sponsor and the Academies' board. They are formally determined in a 'statement of task'.[79]

In the case of the Royal Society, the British counterpart to the National Academies, the connection to the policy process is less directly pronounced with a more general emphasis on scientific 'excellence' and its benefit for humanity.

Dating back to the 1660s, the Royal Society of London is the oldest national academy of science 'in continuous existence' in the world.[80] Its mission is 'to recognise, promote, and support excellence in science and to encourage the development and use of science for the benefit of humanity'.[81] In addressing the highly contested suggestion of climate engineering, the Society's self-proclaimed aim was to present an 'authoritative' assessment, and indeed, the report left an irrefutable mark on the emerging debate over climate engineering, especially in advancing the 'Plan B' narrative.[82]

Finally, in the world of climate policy making, the Intergovernmental Panel on Climate Change (IPCC) takes up a particularly prominent place in providing scientific assessments. In contrast to both the US National Academies and the Royal Society, the IPCC is set up as an international, representative parliamentary body. This formal integration of political representation and scientific assessment has helped the organisation to prominence within science and science-policy studies; it has provided social scientific research with a textbook example of a 'boundary organisation'.[83] The IPCC's so-called Summaries for Policy Makers (SPMs) have been dubbed 'the frontline of negotiations between climate science and climate policy' in this sense.[84] The organisation is, as Spencer Weart put it, 'neither a strictly scientific nor a strictly political body, but a unique hybrid'.[85] And what is more, in contrast to the National Academies, the IPCC is a specialist organisation, focusing specifically on the issue of anthropogenic climate change.

Selection of experts via the congressional inquiry

In contrast to the scientific assessments, the focus of the kind of advice that is provided upon congressional inquiry is more directly policy oriented. Congressional inquiries, including hearings as well as legislative assessments, provide a critical political platform to establish newly emerging topics, as we have seen in Chapter 2: they are an arena of 'purposive' communication as we saw earlier.[86]

Hearings and legislative reports are forms of expert advice which are set up *within* the legislative branch, primarily catering to the needs of Congress. They provide the opportunity to invite pertinent experts from beyond the federal bureaucracy to determine critical evidence on issues of policy concern. It is the policymakers who get to select witnesses or task congressional research agencies with the provision of advice. The selection of expert witnesses is therefore a highly strategic undertaking which primarily furthers particular political goals.[87] Expert testimonies thus can be understood as politically mediated expert observations. They appear as a kind of 'ideologically and politically reliable' kind of expertise, according to Sabine Maasen and Peter Weingart.[88]

Similarly, the Congressional Research Service (CRS) or the Government Accountability Office (GAO) are congressional support bodies which provide information and research needs directly to Congress. The Research Service dates to an initiative from 1914, which was driven by progressive-era ideals, emphasising the relevance of the 'acquisition of knowledge for an informed and independent legislature'.[89] The initiative first led to the establishment of a Legislative Reference Service, which was then turned into the Congressional Research Service (CRS) in 1970. Both the Government Accountability Office and the Research Service are research support agencies, and as such, are established to provide nonpartisan and objective expertise. CRS in this sense institutionalises a clear division between evidence base and policy making. According to CRS itself, the agency

> makes no legislative or other policy recommendations to Congress; its responsibility is to ensure that Members of the House and Senate have available the best possible information and analysis on which to base the policy decisions the American people have elected them to make.[90]

The Government Accountability Office (GAO) in contrast specialises more specifically on government oversight and auditing services for Congress.[91] It provides Congress with reports and analyses of activities within the executive branch.

If we take a closer look at the invited experts that the congressional inquiry on climate engineering has rested on, it is the decided focus on scientists that sticks out. Table 7.1 suggests that it is primarily scientists – and academic scientists at that – who have most prominently been selected as congressional experts on the issue of climate engineering. Roughly half the experts who have been selected to inform the congressional inquiry have also contributed to one of the examined scientific assessment reports. This is also reinforced by GAO's declaration that the agency selected relevant experts based on their 'participation on a geoengineering panel, the number of articles authored in peer-reviewed literature, and recommendations from other experts'.[92] With regards to their institutional background, the congressional experts who have been selected to inform an official political record on climate engineering thus stand in stark contrast to those expert witnesses who brought the issue onto the congressional agenda before this official inquiry. As we have seen in Chapter 5, these 'masked agenda setters' primarily came from think tanks instead of universities and research institutes.

This decided focus on scientific (and even academic) selection criteria seems remarkable if we consider the fact that the experts were invited to inform a policy agenda on climate engineering and not a scientific argument. This means that scientific experts appear here to not merely speak on the science; they are invited to define political stakes, envision feasible futures and 'to decide public meanings'.[93] The apparent lack of alternative perspectives or competing forms of invited expert voices suggests that, in this case of climate engineering, policymakers see scientists as the respectively crucial experts, as the fact bearers, and as the critical 'problem solvers'.[94] These congressional experts become politically relevant for the issue at hand precisely in their role *as prominent researchers*. This means that they claim epistemic authority on climate engineering in both science and politics; they connect scientific with political visibility.

The literature further complements this picture, adding another realm to the epistemic authority of these scientific experts – namely the media. Holly Buck, for example, found that between 1990 and 2010, '36 per cent of assertions made in the media' about climate engineering were presented by only nine scientists.[95] And seven of those nine scientists have also been invited to inform the congressional inquiry on climate engineering. These findings thus emphasise just how small the world of climate engineering was in the early 2000s.[96] Eli Kintisch had coined the term 'geoclique' for these experts to stress that the issue of climate engineering – its research, presentation in the media, and assessment in politics – was essentially defined by a small group of individuals. The congressionally selected experts thus occupy a particularly prominent and powerful position. Not only are they structuring and informing the US political inquiry into climate engineering, but they also shape the scientific debate and the discussion on the issue in the media.

Against this backdrop, we can revisit Ann Keller's observation that 'there are strong disincentives for academic scientists to testify before Congress', mainly because of the risk of being perceived as politically biased.[97] The selection of expert witnesses who took the stage in this arena of 'staged advice' paints a different picture. It suggests that scientists are able to reinforce their epistemic authority precisely by formulating distinct policy issues in scientific terms. Even academic scientists might effectively utilise congressional hearings to advance the relevance of their research in defining or addressing societal challenges and potential response measures.[98] We can observe a similar dynamic in the early politicisation of climate change and biodiversity loss (see Chapter 4).[99] There is hardly any account of the politicisation of climate change that does not build on the driving role of either individual experts, groups, or networks of scientists – 'non-sinister conspiracies', 'rainforest mafias', 'geocliques' – that pushed the issue into the political realm. Scientists appear in these cases as spokespeople, bringing issues from science to the attention of politics.

This, in turn, raises a question for future research: namely, how emerging political agendas might influence inner scientific differentiation. How, in other words, does reference to political agendas structure the scientific selection of research topics or the formulation of emerging research programs? Many of the relevant climate engineering experts, for example, emphasise that they have initially worked on the issue of climate engineering in their 'spare time'.

To sum up, 'staged advice' provides an expert setting which channels scientific expertise from academic research contexts into the political system. Although the two channels of scientific assessments and congressional inquiry do in fact entail a respectively different selection of experts, both have effectively selected and invited primarily academic scientists to establish an official record on climate engineering. The experts that take stage in this arena of staged advice can claim political visibility precisely based on their scientific, even academic credentials.

DEVISING A PROJECT OF CLIMATOLOGICAL CULTIVATION AND CONTROL

Their political authority in co-defining the stakes of climate engineering rests on their scientific authority.

Science for national needs

Beyond this publicly orchestrated layer of staged advice, we have seen that scientific expertise came to shape the emerging politics of climate engineering from within the federal bureaucracy. Acronyms such as NASA, NOAA, EPA, NCAR, NSF, but also DOE, USDA, DOD, DOS or USGS provide central nodes via which the political system began internalising climate engineering into the federal bureaucracy (as we have seen in Chapter 2).[100] Specifically, these acronyms stand for scientific and regulatory bodies within the federal bureaucracy. They institutionalise original research capacities within the state, both within particular departments and via independent agencies. This means that scientific expertise connects to the political realm in this context not by being externally invited to provide scientific advice, but by internalising the research capacities themselves.

These agencies, then, can be understood as the targeted and formal organisation of 'science for national needs'.[101] They essentially gear research capacities (more or less formally) to the national strategic goals of the federal government. Their research programs are aligned to political priorities, for example, via funding structures, such as executive and congressional budget decisions, or via organisational and inter-agency program missions. The specific scope and overall direction of their research is thus subject to congressional politics and executive budget decisions.

These agencies became relevant in a notably different capacity than the invited experts who provided what I have called, 'staged advice'.[102] Instead of shaping the emerging politics of climate engineering by actively framing the debate and official record on the issue, these expert agencies seem subject to political efforts to steer the development of climate engineering expert capacities. These agencies were tasked with addressing the technicalities established by the official record as the recipients of federal funds, charged with advancing research on the issue, or with devising a regulatory infrastructure. These agencies thus reflect how politics has structurally internalised climate engineering; they suggest how politics has translated and adopted this issue and made it legible to national strategic concerns.

In the following, we will take a closer look at how precisely these expert agencies have articulated and given concrete shape to this notion of engineering the climate within the United States. I focus here on five agencies: the National Science Foundation, the National Oceanic and Atmospheric Administration (NOAA), the National Aeronautics and Space Administration (NASA), the Environmental Protection Agency (EPA), and the Department of Energy.

The National Science Foundation

The National Science Foundation (NSF), to begin with, is the US government's central agency for supporting basic research. In contrast to the other agencies that we will turn to in the following, NSF has advanced climate engineering research by providing grants, not by hosting in-house research or governing particular measures. Climate engineering thus appears in this context as an emerging topic of research. NSF was seen to play a critical role in advancing both social scientific, as well as natural scientific research on climate engineering. According to the Science Committee, the agency has essential expertise for guiding questions into 'domestic and international governance, economics, and risk assessment and management', as well as 'ethical considerations'.[103] Experts and policymakers therefore agreed that, on the one hand, NSF should contribute to 'informing public engagement strategies' around climate engineering.[104] On the other hand, based on its observational and modelling expertise, the Science Committee suggested that the agency should continue playing a leading role in supporting solar radiation management research.[105] The agency has already funded critical modelling studies, for example, conducted by Alan Robock and others at Rutgers University.[106]

The National Oceanic and Atmospheric Administration

The National Oceanic and Atmospheric Administration (NOAA) is a scientific agency within the Department of Commerce. Together with the Department of

Energy, NOAA has been one of the most talked about, referenced, and prominent expert agencies within the US political exploration of climate engineering.[107] The very idea of climate engineering crucially rests on the modes of observation that NOAA provides.

Through the lens of NOAA, anthropogenic climate change primarily emerges as an issue of ever more precise climatological observation and measurements. The agency makes this issue legible in the motion of particles, in air and ocean temperatures, in vegetation or atmospheric humidity. Since its inception in 1970, NOAA has specialised in precise observations and 'impeccable measurements' of these data points.[108] The first report of what is now called NOAA's Global Monitoring Division celebrated the historical importance of this mission. The authors argued that this report 'has as its genesis the unknown observer of antiquity who, realising that an observation unrecorded was an observation lost, inscribed a record on stone or clay'.[109] In the eyes of experts and policymakers alike, this climatological observational infrastructure and expertise made NOAA the logical home for any efforts to advance a coordinated agenda on climate engineering.[110] The Science Committee, especially, highlighted NOAA's 'ground-truthing equipment and software' in this context.[111] This kind of observational expertise was deemed relevant to any approach that would involve 'chemical inputs into the environment that would directly or indirectly impact ocean waters, e.g. stratospheric sulfate injections and ocean fertilization'.[112]

The organisational precursors of NOAA date all the way back to the very beginning of the United States of America. The agency was formed through the integration of 'America's first physical science agency, America's first agency dedicated specifically to the atmospheric sciences, and America's first conservation agency'.[113] This namely includes the US Coast and Geodetic Survey (initiated in 1807 by President Thomas Jefferson), the Weather Bureau (founded in 1870), and the US Commission on Fish and Fisheries (established in 1871). To quote Zeke Baker once again, NOAA's organisational history illustrates how the agency has persistently matched scientific challenges of precise climatological measurement to shifting political agendas over the years – from commercial and agricultural challenges to the challenge of the ozone hole, to global warming, to assessing the potential of fertilising the oceans or reducing incoming sunlight. NOAA's history connects the project of climate engineering to a long line of different science-state configurations in which precise climatological observation became of direct political relevance. NOAA's history thus suggests how intricately interwoven knowledge production and problem observation, or the 'discovery' and the tackling of climate change is.

The National Aeronautics and Space Administration

The National Aeronautics and Space Administration (NASA) implies a comparable, yet different, institutional home for the pursuit of climate engineering. NASA's mission had been grand from the very start.[114] The agency was born in the midst of the Cold War, charged with rising to the challenge of ensuring global pre-eminence by means of science. The decision to create NASA was driven by the launch of Sputnik in October 1957. According to a former chief historian at NASA, 'a country that aspired to global pre-eminence could not let that challenge pass'.[115] In July 1958, President Eisenhower signed the National Aeronautics and Space Act into law, formally establishing NASA. Apart from its more direct space-related remits, the final Space Act lists 'the expansion of human knowledge of phenomena in the atmosphere and space' as a core task of NASA.[116]

As in the case of NOAA, NASA's critical role in providing policy-relevant climate change expertise, too, was articulated and institutionalised in the formation of the US Global Change Research Program and its predecessors. Both agencies work closely together in many respects yet follow a rather clear division of labour. For example, while NASA's satellites provide extensive observational coverage, NOAA's Earth-based observations provide precision where needed. While NASA focuses on experimentation and technology development, NOAA ensures operational continuity. It was the combination of these foci which made NASA relevant to political efforts in terms of developing climate engineering expert capacities. Like the case of NOAA, policymakers and experts agreed that the agency's observational expertise and its airborne and satellite-based monitoring capacities would be critical not only for studying solar radiation management, but also for monitoring and managing any such scheme, once deployed and in place.[117]

Through the lens of NASA, tackling climate change not only emerges as a challenge to observational precision, but as a task of properly monitoring and managing particles. It also appears as a grand project of mind-boggling, techno-scientific innovation. Climate engineering came on NASA's radar through its innovation-driven mission. The agency had explored 'the practicality of using a solar shield in space to deflect sunlight and reduce global temperatures' through its former independent Institute for Advanced Concepts (NIAC).[118] NASA's goal with NIAC was to foster 'innovation that stretched the imagination of the technical community and encouraged revolutionary creativity'.[119] The program was terminated in 2007. In contrast to NOAA, NASA thus represents a much more public relations-oriented national vision of technological innovation; its tasks, missions, and programs are selected very strategically. The fact that NASA has not committed to pushing climate engineering more publicly might thus be read as suggesting just how controversial the issue remains until today.

The Environmental Protection Agency

In contrast to both NOAA and NASA, the Environmental Protection Agency (EPA) became relevant to the emerging politics of climate engineering via its regulatory and governance capacity. Tackling climate change appears here as an issue of environmental safeguarding. The agency had been established by President Nixon together with NOAA in 1970 in an effort to respond to rising pressure from a growing environmental movement (see also Chapter 4).[120]

In the context of the climate engineering 'inventory' of 2009 and 2010, EPA appeared as one of the few agencies with expertise and authority to regulate and govern any emerging climate engineering activities.[121] The House Science Committee therefore deemed the EPA as imperatively involved in the regulation of any testing or deployment of climate engineering activities. And indeed, EPA has already 'initiated rulemakings to regulate certain geoengineering activities'.[122] In 2010, for example, the agency established a rule that qualifies the technical requirements necessary for long-term capture and storage of CO_2 in injection wells to ensure safe drinking water.[123] And five years later, EPA put forth emission guidelines in the form of a Clean Power Plan,[124] which was proposed to be established under the Clean Air Act.[125] Paralleling the American Clean Energy and Security Act of 2009, which explicitly disqualified ocean fertilisation approaches as a legitimate form of carbon sequestration,[126] the EPA explicitly ruled direct air capture (DAC) measures as a *non-option* for counteracting climate change. In its final rule on Carbon Pollution Emission Guidelines, the EPA listed DAC technologies as an example 'of measures that may not be counted toward meeting a CO_2 emission performance level'.[127]

Through this lens of the EPA, tackling climate change becomes an issue of environmental safeguarding. As a potential response measure, climate engineering is thus observed and regulated according to its environmental impacts and its pollution and health hazards.

US Department for Energy

Finally, the US Department of Energy (DOE) emerged as a critical home for climate engineering within the federal bureaucracy, particularly in the context of carbon dioxide removal measures. The House Science Committee envisioned that the DOE 'should lead any federal research program into air capture and non-traditional carbon sequestration'.[128] As we have seen in Chapters 2 and 5, over the first two decades of this millennium, Congress tasked the department with 'kickstarting' carbon dioxide removal and especially direct air capture technology. Via the DOE, the political system sought to steer not only research, but technology development and commercialisation in a number of ways and formats – from research and development programs to cash prize initiatives and direct investments, to demonstration facilities. Apart from these energy-related climate engineering approaches, DOE has supported solar radiation management research through both its Sandia National Laboratories and Pacific Northwest National Laboratory.[129] The Congressional Research Service (CRS) pointed out that such a diverse involvement in climate engineering activities might lead to conflicts of interest along the way.[130]

The agency makes climate change politically legible as a technological issue to effective energy generation. Climate engineering – and especially Carbon Removal – appears as a critical answer to this issue. Through the lens of DOE, climate engineering is thus less the result of an agenda of better climatological understanding, of measuring, monitoring, and eventually managing climatological dynamics, as is the case for NOAA and NASA. In this case of DOE, climate engineering rather emerges in its engineering dimension. Over the years, the department politically internalised and institutionalised climate engineering as a critical building block of national 'clean' power generation and climate change technology development.[131]

To sum up, this section has anchored the emerging politics of climate engineering to what I suggested calling a corresponding *expert infrastructure*. It has shed light on the various expert settings via which science and politics came together in shaping the career of climate engineering in the United States and formulated this project of climatological cultivation and control. By illustrating these expert settings as critical components of the climate engineering expert infrastructure, the section qualified different forms in which scientific research and political agendas are reciprocally coupled. Not only do the congressional experts and expert agencies inform the political process, but conversely, it is also this political process that 'makes' these researchers, agencies, and departments into experts, consequently shaping research agendas in one way or another.

Returning to Grundmann's point that expertise is always delivered 'at the request of someone else'[132], this section illustrated just how complex this notion of 'request' becomes empirically. We have seen how differently these expert voices connect to politics, that is, how differently scientific expertise becomes relevant to the political process. And we have seen how these expert voices point us to very different, even conflicting, timescales in the trajectory of climate engineering. Congress provides a relevant platform for the politicisation of issues by inviting scientific experts to set the agenda on controversial issues – such experts then formulate, qualify, and contest climate engineering as a legitimate policy measure. Federal agencies, in contrast, stabilise the generation of problem-relevant expertise within the state. As a result, climate engineering appears as a new and contested issue through the lens of congressional politics, while through the lens of federal agencies, climate engineering appears as merely another chapter in the much longer-standing history of the federal institutionalisation of climate relevant expertise (particularly in the form of the US Global Change Research Program). Both conflicting temporalities are relevant to an understanding of how this notion of climate engineering became 'serious politics'. And the picture would get even more complicated if we systematically considered the scientific papers that have structured the techno-scientific struggles, described in Chapter 2. We cannot make sense of the interrelation of science and politics in climate engineering by merely studying the provision of expert advice on a supposedly new and controversial topic on the political agenda. Rather, this expert infrastructure suggests how the emergence of political issues and their respective response measures are bound up with the emergence of research agendas and expert perspectives.

NOTES

1 Kintisch (2010).
2 Fleming (2010: 253).
3 Roman and Carson (2009: 55).
4 Turner and Isenberg (2018: 181).
5 See, e.g., Kincaid and Roberts (2013); Turner and Isenberg (2018: 180f.).
6 See, e.g., Roman and Carson (2009); Kincaid and Roberts (2013); Chait (2013); Turner and Isenberg (2018).
7 See, e.g., Figure 4–4 in Turner and Isenberg (2018: 185).
8 Notably, this was despite the fact that the administration had pushed some relevant climate legislation. Kincaid and Roberts (2013). See also, e.g., Roman and Carson (2009); Kincaid and Roberts (2013); Chait (2013); Turner and Isenberg (2018).
9 Kincaid and Roberts (2013: 47).
10 Kincaid and Roberts (2013: 48).
11 Political attention to ocean fertilisation seemed largely favourable during these years (see, e.g., US House of Representatives, Select Committee on Energy Independence and Global Warming (2007: 95); US Senate, Subcommittee on Oceans, Atmosphere, Fisheries, and Coast Guard of the Committee on Commerce, Science, and Transportation (2007); S.2307 (2008: 20ff.); US Senate, Subcommittee on Energy and Water Development of the Committee on Appropriations (2009: 42);

US House of Representatives, Subcommittee on Commerce, Justice, Science, and Related Agencies of the Committee on Appropriations (2009: 304); US House of Representatives, 111th Congress (2009: 33)). In 2009, however, the US House of Representatives passed the American Clean Energy and Security Act of 2009 (not enacted), which sought to implement a so-called cap and trade system in the United States (H.R.2454 (2009)). And while the proposed bill called for a national Carbon Sequestration strategy which would have included climate engineering approaches – sequestration is defined in the text as 'the separation, isolation, or removal of greenhouse gases from the atmosphere [...]' – the bill explicitly excluded ocean fertilisation measures (H.R.2454 (2009: 1389)). The Senate version of this bill ultimately failed 'due to lack of support and political capital' (Kincaid and Roberts (2013: 44)).

12 S.2307 (2008: 5).
13 S.2307 (2008: 2).
14 S.2307 (2008: 5–6).
15 S.2307 (2008: 20ff.). The original bill, as introduced by John Kerry on 5 November 2007 did not yet contain any reference to ocean fertilisation; this was only added in the reported 2008 version of the bill.
16 See also Mitchell (2002).
17 US House of Representatives, Committee on Science and Technology (2009: 35).
18 Scott (1998).
19 Some of the findings from this section have been published in Schubert (2019, 2021).
20 See, e.g., Sismondo (1999); Edwards (2010); Gramelsberger and Feichter (2011: 9–90).
21 See, e.g., Eule in US House of Representatives, Committee on Government Reform (2006: 75); Holdren in US Senate, Committee on Commerce, Science and Transportation (2009: 13f.).
22 Allan (2017: 137).
23 Allan (2017: 138).
24 Eule in US House of Representatives, Committee on Government Reform (2006: 9); Schnare in US Senate, Committee on Environment and Public Works (2007: 157); US House of Representatives, Committee on Science and Technology (2009: 39, 110, 154, 163, 300, 345); Lubchenco in US House of Representatives, Committee on Science and Technology (2010a: 79); US House of Representatives, Committee on Science and Technology (2010b: 5, 7).
25 See, e.g., Lawrence and Crutzen in Blackstock and Low (2019: 92). In the United States, the most prominent of these community-spanning modelling projects is the Geoengineering Model Intercomparison Project (GeoMIP) (see, e.g., Kravitz,

Robock, and others (2011); Kravitz, Caldeira, Boucher, and others (2013); Tilmes and others (2013); Kravitz, Robock, Tilmes, and others (2015); Kravitz, MacMartin, Visioni, and others (2020)). For a list of all simulations and publications that either come from the GeoMIP working group or use their models, see the GeoMIP website. In Europe, the EU-funded project on Implications and Risks of Engineering Solar Radiation to Limit Climate Change (IMPLICC), which ended in 2012, has received much attention (Implications and Risks of Engineering Solar Radiation to Limit Climate Change 2019).

26 Experts also refer to less 'natural' analogues to communicate climate engineering, e.g., cloud formation in ship tracks (see, e.g., Lee and Rasch in US House of Representatives, Committee on Science and Technology (2009: 35, 160)).

27 Carlin in US Senate, Committee on Environment and Public Works (2007: 126).

28 Robock in US House of Representatives, Committee on Science and Technology (2009: 49, 51).

29 US House of Representatives, Committee on Science and Technology (2009: 6).

30 See, e.g., Caldeira, Lane, Sheperd, Keith, and Morgan in US House of Representatives, Committee on Science and Technology (2009: 24, 25, 35, 80, 148, 284); Schnare in US Senate, Committee on Environment and Public Works (2007: 67); US Senate, Subcommittee on Energy and Water Development of the Committee on Appropriations (2009: 42); US House of Representatives, Committee on Science and Technology (2010b: 4).

31 Lackner in US House of Representatives, Committee on Science and Technology (2009: 172); US Environmental Protection Agency EPA (2010: 77234).

32 Schnare in US Senate, Committee on Environment and Public Works (2007: 155); see also George in US House of Representatives, Select Committee on Energy Independence and Global Warming (2007: 95).

33 Robock et al. in US House of Representatives, Committee on Science and Technology (2009: 59–60).

34 This is how the economist Alan Carlin put it, as quoted by one expert witness in 2007 (Carlin qtd. by Schnare in US Senate, Committee on Environment and Public Works (2007: 126)).

35 Schnare in US Senate, Committee on Environment and Public Works (2007: 64).

36 See, e.g., Schnare in US Senate, Committee on Environment and Public Works (2007: 64, 68, 110, 126, 155); US Senate, Subcommittee on Energy and Water Development of the Committee on Appropriations (2009: 42).

37 Caldeira in US House of Representatives, Committee on Science and Technology (2009: 20). See also: US House of Representatives, Committee on Science and Technology (2009: 152, 157, 309); Washington Post as qtd. in US Senate, Committee on Environment and Public Works (2015: 10).

DEVISING A PROJECT OF CLIMATOLOGICAL CULTIVATION AND CONTROL

38 Caldeira in US House of Representatives, Committee on Science and Technology (2009: 16). See also Caldeira, Rasch, Long in US House of Representatives, Committee on Science and Technology (2009: 20, 152, 300); US Senate, Committee on Environment and Public Works (2015: 10); US Environmental Protection Agency EPA (2010: 77234). For a critical position, see Chisholm et al. qtd by Romm in US House of Representatives, Select Committee on Energy Independence and Global Warming (2007: 39f.).
39 Schnare in US Senate, Committee on Environment and Public Works (2007: 68, emphasis added); see also Schnare in US Senate, Committee on Environment and Public Works (2007: 111); for the case of CDR, see, e.g., US Senate, Subcommittee on Energy and Water Development of the Committee on Appropriations (2009: 42).
40 Carlin qtd. by Schnare in US Senate, Committee on Environment and Public Works (2007: 126, 156).
41 Lackner in US House of Representatives, Committee on Science and Technology (2009: 172).
42 Caldeira in US House of Representatives, Committee on Science and Technology (2009: 16; see also: 20, 22, 24, 35, 90, 300, etc.). US House of Representatives, Select Committee on Energy Independence and Global Warming (2007: 40); Schnare in US Senate, Committee on Environment and Public Works (2007: 73, 111, 112).
43 US House of Representatives, Committee on Science and Technology (2010b: 23).
44 Schneider in US Senate, Committee on Environment and Public Works (1997: 122).
45 Rasch in US House of Representatives, Committee on Science and Technology (2009: 152, emphasis added).
46 Robock, Caldeira in US House of Representatives, Committee on Science and Technology (2009: 45, 90, 109).
47 Long in US House of Representatives, Committee on Science and Technology (2009: 301).
48 Robock, Rasch, US Government Accountability Office in US House of Representatives, Committee on Science and Technology (2009: 38, 122, 159).
49 Rasch in US House of Representatives, Committee on Science and Technology (2009: 152).
50 Robock in US House of Representatives, Committee on Science and Technology (2009: 44); see also Wigley qtd. by Schnare in US Senate, Committee on Environment and Public Works (2007: 111f.).
51 Rasch in US House of Representatives, Committee on Science and Technology (2009: 157, emphasis added).
52 Robock in US House of Representatives, Committee on Science and Technology (2009: 49, emphasis added). See also US House of Representatives, Committee on

Science and Technology (2010b: 29); Robock, Long, Caldeira, Rasch in US House of Representatives, Committee on Science and Technology (2009: 44, 52, 82, 90, 121, 311).
53 Sismondo (1999: 247).
54 Sismondo (1999: 256).
55 US Senate, Committee on Commerce, Science and Transportation (2009: 39ff.); US House of Representatives, Committee on Science and Technology (2009: 7, 32, 39, 45, 47f., 82, 110, 112f., 118ff., 149, 152, 158f. 285); US House of Representatives, Committee on Science and Technology (2010b: 7); Lubchenco in US House of Representatives, Committee on Science and Technology (2010a: 79); US Senate, Committee on Environment and Public Works (2013: 25); US House of Representatives, Subcommittee on Africa and Global Health of the Committee on Foreign Affairs (2010: 72ff.).
56 See, e.g., Rasch, Lackner, Robock, Caldeira in US House of Representatives, Committee on Science and Technology (2009: 161, 175, 118, 211). Pressing scientific and engineering challenges concern, e.g., particle formation and evolution. Rasch, for example, describes different types of models that would be necessary to inform such an undertaking in the future (in US House of Representatives, Committee on Science and Technology (2009: 158)).
57 Robock in US House of Representatives, Committee on Science and Technology (2009: 50, see also: 7, 48).
58 'Fingerprinting – detection and attribution of human intervention effects on climate – must be an important area for research if we are to be able to conduct adaptive and successful management of geoengineering. As this topic is closely interconnected to basic climate science, the program to extend research into intentional intervention should belong in the US Climate Science Program' (Long in US House of Representatives, Committee on Science and Technology (2009: 308); see also US House of Representatives, Committee on Science and Technology (2009: 45f., 50ff., 123, 141, 307).
59 See, e.g., Green in US House of Representatives, Select Committee on Energy Independence and Global Warming (2010: 68); US House of Representatives, Committee on Science and Technology (2009: 34f., 145, 183, 279, 317).
60 US House of Representatives, Committee on Science and Technology (2009: 186).
61 US House of Representatives, Committee on Science and Technology (2009: 16).
62 US House of Representatives, Committee on Science and Technology (2009: 24).
63 Caldeira in US House of Representatives, Committee on Science and Technology (2009: 17). US House of Representatives, Committee on Science and Technology (2009: 19ff., 36, 168f.); S.2744 (2009: 2); US Senate, 112th Congress (2011: 3).

64 US House of Representatives, Committee on Science and Technology (2009: 7; see also 31, 34f., 116, 222).
65 US House of Representatives, Committee on Science and Technology (2010b: III).
66 US House of Representatives, Committee on Science and Technology (2009: 8–10); see also, e.g., US Senate, Committee on Environment and Public Works (2015: 12); US House of Representatives, Committee on Science and Technology (2010b: III); US House of Representatives, Committee on Science and Technology (2009: 5, 255, 360, 364); Royal Society (2009: xiii).
67 These findings speak to observations from the literature which suggest that social scientific expertise figures prominently in technology policy processes mostly when it adds to narratives of techno-scientific control, not when it questions them (see, particularly, Smallman (2020)).
68 US House of Representatives, Committee on Science and Technology (2009: 47).
69 Robock in US House of Representatives, Committee on Science and Technology (2009: 44); see also, e.g., US House of Representatives, Committee on Science and Technology (2009: 8f.).
70 US House of Representatives, Committee on Science and Technology (2010b: 11); see also, e.g., US House of Representatives, Committee on Science and Technology (2009: 28, 72, 361).
71 Sheperd in US House of Representatives, Committee on Science and Technology (2009: 28, emphasis added).
72 Critical positions on climate engineering, in contrast, hardly seem to follow disciplinary boundaries. Meanwhile, Jim Fleming, Stephen Schneider, or (to some extent) Alan Robock, for example, provide rather sceptical accounts (Kellogg and Schneider (1974); Schneider (1996); Fleming (2006); Robock (2008); Fleming (2010). Scott Barrett and Lee Lane pursue distinctly favourable positions (Barrett 2008; Bickel and Lane (2009)).
73 Grundmann (2017: 26).
74 For an instructive summary of these observations, see Eyal (2019: 105f.).
75 See also Hilgartner (2000).
76 Gupta and Möller (2019: 481).
77 US National Academies of Sciences (2017a). See also Hilgartner (2000: 21 ff.) for an instructive account on the role of the Academy within the US state.
78 The two volumes on 'climate intervention' measures, for example, have been sponsored by the Department of Energy, the National Aeronautics and Space Administration (NASA), the National Oceanic and Atmospheric Administration (NOAA), the Arthur L. Day Fund, and 'the intelligence community' (US National Research Council (2015a: ii)).

ENGINEERING THE CLIMATE

79 US National Academies of Sciences (2017b).
80 Royal Society (2017a).
81 Royal Society (2017b).
82 Royal Society (2009: v); see also the testimony of John Shepherd in US House of Representatives, Committee on Science and Technology (2009: 27).
83 See, e.g., Beck (2009, 2011, 2016); Jobst (2010); Miller (2001); Poloni (2009).
84 Stilgoe (2015: 24). While climate engineering approaches remained rather peripheral for a while, they have been steadily considered in the IPCC's cyclical assessment reports throughout the years (see, e.g., Petersen (2014) for a general account of the emergence of Geoengineering within the IPCC). In the first assessment report (FAR) from 1990, suggestions for technical climate intervention were limited to large-scale afforestation projects (Intergovernmental Panel on Climate Change (1990: 287, 301)). In the IPCC's second assessment report (SAR) from 1995, 'geoengineering options' were included as a means 'to counterbalance greenhouse-gas induced climate change' (Intergovernmental Panel on Climate Change (1995: 18, 51)). Notably, this report already applied the definition that would eventually prevail in the successful establishment of these measures in US climate policy during the early 2000s. It presented climate engineering measures as potential 'last resort options for the future' that should be kept 'in reserve in case of unexpectedly rapid climatic change' (Intergovernmental Panel on Climate Change (1995: 90, 802)). In 2001, the third assessment report (TAR) called for 'basic inquiry in the area of geo-engineering' as a means for addressing current knowledge gaps (Intergovernmental Panel on Climate Change (2001: 13)). In contrast to the previous report, it focused exclusively on measures of carbon dioxide removal. The fourth assessment report (AR4), Working Group III (Mitigation of Climate Change), considered both ocean fertilisation measures and solar radiation management, yet argued that both measures remained 'largely speculative and unproven' and that 'reliable cost estimates for these options have not been published' (Intergovernmental Panel on Climate Change (2007: 15)). Similarly, the fifth assessment (AR5) report pointed to the 'limited evidence' both on solar radiation management and carbon dioxide removal measures and emphasised the anticipated risks of such an approach to counteracting climate change – this time, in the Summary for Policy Makers (SPM) by Working Group I (The Physical Science Base) (Intergovernmental Panel on Climate Change (2013: 29)).
85 Weart (2008: 153).
86 Keller (2009: 95).
87 Bimber (1996: 21); Keller (2009: 96).
88 Maasen and Weingart (2006: 6).
89 Brudnick (2008: 1).
90 Brudnick (2008: 2).

91 See, e.g., Kaiser (2007).
92 US Government Accountability Office (2010b: 4).
93 Smallman (2020: 591).
94 Smallman (2020: 593).
95 Buck quoted by Hamilton (2013: 219).
96 See, e.g., Hamilton (2013: 72); Stilgoe (2015: 186); Morton (2016: 157, 333). For a detailed analysis, see 'Mapping the Landscape of Climate Engineering' between 1971 and 2013, see Oldham and others (2014).
97 Keller (2009: 97).
98 For the controversial nature of this argument, see, particularly, the account by Sarewitz and Pielke (2007: 10f.).
99 Hart and Victor (1993); Hannigan (2006: 157f.).
100 US House of Representatives, Committee on Science and Technology (2009: 5, 48, 54, 123, 172, 263ff.); US House of Representatives, Committee on Science and Technology (2010b: 28f.).
101 Mukerji (2014). With this concept, Mukerji draws attention to the fact that by funding scientific research in various forms, the government maintains a pool of relevant expertise as a kind of reserve of problem-solvers 'if push came to shove' (Mukerji (2014: 66)).
102 Of course, there is some overlap in their appearance: representatives of these expert agencies have also shaped the politics of climate engineering by actively framing the stakes of the debate, for example, by testifying before Congress and contributing to assessment reports. But more importantly than providing 'staged advice', these agencies suggest how climate engineering expert capacities have been built up within the federal bureaucracy over the years.
103 US House of Representatives, Committee on Science and Technology (2010b: 10).
104 US House of Representatives, Committee on Science and Technology (2010b: 11).
105 See, e.g., US House of Representatives, Committee on Science and Technology (2009: 81f., 114, 120, 149).
106 Alan Robock's research on SRM at Rutgers University, for example, was supported by a grant on 'Collaborative Research in Evaluation of Suggestions to Geoengineer the Climate System Using Stratospheric Aerosols and Sun Shading', (February 1, 2008 – January 31, 2011, $554,429) (Robock in US House of Representatives, Committee on Science and Technology (2009: 50, see also 223, 263, 272)).
107 See Introduction and Chapter 1 for concrete examples on how NOAA has emerged since the early 2000s as a central target of political efforts to steer the development of both SRM and CDR expert capacities within the state.

108 National Oceanic and Atmospheric Administration (1974: 2).
109 National Oceanic and Atmospheric Administration (1974: 1).
110 See, e.g., Feely in US Senate, Subcommittee on Oceans, Atmosphere, Fisheries, and Coast Guard of the Committee on Commerce, Science, and Transportation (2007: 83); Solomon in: US House of Representatives, Subcommittee on Commerce, Justice, Science, and Related Agencies of the Committee on Appropriations (2009: 305); Lubchenco in US House of Representatives, Committee on Science and Technology (2010a: 79); US Government Accountability Office (2010b: 20f.); Bracmort and Lattanzio (2013: 28).
111 US House of Representatives, Committee on Science and Technology (2010b: 14, see also: 17); US House of Representatives, Committee on Science and Technology (2010a: 7); US House of Representatives, 112th Congress (2012: 38).
112 US House of Representatives, Committee on Science and Technology (2010b: 16). Particularly the Office of Oceanic and Atmospheric Research (OAR) and the National Environmental Satellite Data and Information Service (NESDIS) were listed as pertinent in this regard (US House of Representatives, Committee on Science and Technology (2010b: 12f.)).
113 National Oceanic and Atmospheric Administration (2017).
114 The agency was equipped with an organisational unit specifically designated to 'documenting and preserving the agency's remarkable history '(National Aeronautics and Space Agency (2017a)). And this division, in turn, has its own organisational history that is documented and preserved.
115 Dick (2008).
116 Dick (2008).
117 See, e.g., Robock in US House of Representatives, Committee on Science and Technology (2009: 91, 120); US House of Representatives, Committee on Science and Technology (2010b: 22, 25). NASA instruments, such as the SAGE series (Stratospheric Aerosol and Gas Experiment), Landsat, or MODIS (Moderate Resolution Imaging Spectroradiometer) were recurrent themes of the congressional inventory of climate engineering relevant expert capacities.
118 US Government Accountability Office in US House of Representatives, Committee on Science and Technology (2009: 263ff.); Bracmort and Lattanzio (2013: 28).
119 National Aeronautics and Space Agency (2017b).
120 Andrews (2010: 227).
121 See, particularly, the GAO report as quoted in US House of Representatives, Committee on Science and Technology (2009: 266); US House of Representatives, Committee on Science and Technology (2010b: 26); Bracmort and Lattanzio (2013: 27f.).

122 Bracmort and Lattanzio (2013: 27).
123 US Environmental Protection Agency EPA (2010). The aforementioned GAO Report (on a coordinated federal strategy on climate engineering) was referenced as the essential legislative history here (US Environmental Protection Agency EPA (2010: 77237)). See also Bracmort and Lattanzio (2013: 12).
124 US Environmental Protection Agency (2015a).
125 US Environmental Protection Agency (2015b: 64966).
126 H.R.2454 (2009: 860).
127 US Environmental Protection Agency (2015a: 64903). This ruling is also reflected in the EPA's proposed federal plan to implement these guidelines. Here, it was stated that emission rate credits (ERC) 'may not be issued to […] direct air capture […]' technologies (US Environmental Protection Agency (2015b: 65094)).
128 US House of Representatives, Committee on Science and Technology (2010b: 22).
129 US House of Representatives, Committee on Science and Technology (2009: 28).
130 Bracmort and Lattanzio (2013: 28).
131 See, e.g., US House of Representatives, Subcommittee on Energy of the Committee on Science (2003); US Department of Energy (2004: 21515); US House of Representatives, Subcommittee on Energy of the Committee on Science (2006); US House of Representatives, Committee on Government Reform (2006); H.R.3607 (2019); US House of Representatives, Select Committee on the Climate Crisis (2020).
132 Grundmann (2017: 26).

CONCLUSION TO PART III

In this third and final part of the book, we have come full circle to where we started at the outset of our analysis. Chapters 5 and 6 traced the career of climate engineering between the beginning and the teens of the new millennium. The story that these two chapters tell is one of a renaissance or re-normalisation of climate engineering within US climate policy. These chapters describe what might be understood as a kind of 'second wave' of visions to intervene in and control the climate. After the politicisation of global warming had drowned out and effectively ended what might be seen as a first wave of hopes to deliberately modify and control the climate (described in *Part II* of this book), these hopes were rekindled at the turn of the new millennium. Chapters 5 and 6 traced how climate engineering incrementally re-gained political traction, this time as a potential response measure to tackling the global societal problem of anthropogenic climate change.

Part III of this book described this incremental re-normalisation of climate engineering as a two-tiered process, structured around the official inquiry into these measures, beginning in November of 2009. Chapter 5 followed the controversial political exploration of these measures before the formal congressional inquiry, focusing in particular on the years from 2003 to 2007. Chapter 6 then zoomed in on the assemblage of climate engineering within and immediately leading up to this formal inquiry, covering the years 2007 to 2016.

Chapter 5 suggested how between 2003 to 2007, climate engineering gained political traction as a kind of techno-fix against climate change. Coinciding roughly with the timeframe of the presidency of George W. Bush, climate engineering popped up in highly controversial debates on the role of technological innovation for tackling the issue of climate change during these years. It moved further into the political limelight here by promising techno-scientific control in the face of this issue. The prospect of climate engineering essentially promised to translate the problem of climate change into a straightforward engineering challenge. It opened a managerial gaze onto the climate change issue, a gaze that problematises climate change not as an issue rooted in techno-scientific intervention, but, to the contrary, as one that calls for more of it. As a result, climate engineering mobilised rather heterogeneous and unlikely climate policy constituencies. It became part of an agenda that sought to push technological innovation as a means to stabilise the political and economic status quo in the face of this issue. It appeared as a tool, catering to national-strategic, as well as economic and corporate concerns in tackling climate change.

Chapter 6 then closed the circle and picked up where we left things off in the first part of this book, namely with the years around the official political inquiry into climate engineering, beginning in November of 2009. The chapter traced the subtle, yet relevant shift in tone that defined this formal inquiry back to the Democratic wave of 2007, when Democratic majorities returned to both chambers of Congress for the first time in twelve years. In contrast to the highly controversial debates during the height of the Bush years, climate engineering was not merely discussed as a techno-fix during this time. The political assessment of these measures was more critical and differentiated, even sceptical of their merit as a policy tool.

Presumably in response to a Democratic constituency that expected policy change on the issue of climate change, policymakers continuously emphasised that climate engineering would not, in fact, provide a solution to the issue at hand. The expected solution was a mitigation of the causes, rather than techno-scientific control over the effects of anthropogenic climate change. Instead of promoting climate engineering as a means of technological progress, the focus shifted to climate engineering as being part of a basic climate science agenda. Climate engineering became part of a climate science agenda during these years that saw climate change as a project of cultivation by human rationality, ingenuity, and reason.[1] Despite being framed as a 'bad idea', these measures became established and institutionalised as a project of climatological cultivation and control.

We saw how science and politics came together in formulating this project of climatological cultivation and control as the chapter zoomed into the kinds of expertise that informed the political exploration of climate engineering. It examined both the defining expert modes of observation, as well as the expert infrastructure that was essential in envisioning and assembling this project. This perspective demonstrated just how deeply interrelated our understanding of problem and response are. In other words, it showed how the rise of climate engineering as a potential policy measure rests on and is embedded in particular modes of observing and problematising the issue of climate change.

Part III of this book thus carved out how the re-normalisation of climate engineering corresponded, once again, to a general shift in the status of climate science for the state. Climate engineering regained political traction in the early 2000s as climate science evolved from a problem-defining to a problem-addressing authority. Climate science, in other words, was not only envisioned as a critical means to understand and decipher, but also to effectively manage and counteract the problem of global warming in this context.

Following this line of reasoning, I suggested that the recent renaissance or re-normalisation of climate engineering can be understood as a kind of *synthesis* which reconciles two historically conflicting roles of climate science within the state. On the one hand, this renaissance was driven by the hope of political control via scientific expertise. Climate science emerged here – once again – as a critical tool at the hands of the state, somewhat mirroring its status from the first half of the twentieth century. This was more blatantly advertised during the Bush administration but continued to play a critical role in US political assessments of climate engineering during the Obama administration. We also currently see it sprouting up again under the Biden administration.

On the other hand, the current politicisation of climate engineering as a 'last resort' or 'Plan B' or 'bad idea' marks an awareness of the limits of such hopes of control, connecting to 'green' notions of environmental safeguarding. The fact that climate engineering has not resonated in recent years as a positive vision of socio-technical innovation points us to the particular societal context in which climate change has been politicised as a societal issue. We get a sense that climate change is not merely an issue of a warming world, but one that marks the limits of control and the potentially detrimental side effects of

CONCLUSION TO PART III

human interventions in its natural environment. As a techno-political project, climate engineering therefore questions established categories of climate policy programs and seems to forge new kinds of alliances between climate science and politics. Particularly in the United States where climate change has grown into the epitome of partisan issues, climate engineering thus promises to shake things up.[2]

Since this is an ongoing and highly dynamic debate, I want to end this Part III of my analysis by daring a glance into the future. To get a sense of where we might be headed, I want to point to three rather different versions of the above-mentioned synthesis of climate science and the state that are looming in this recent debate over climate engineering.

To begin with, we can observe efforts to alter the last resort or 'bad idea' framing of climate engineering in recent years to emphasise the redemptive power of climate science as a tool at the service of humankind. Apart from invoking sheer urgency in tackling the daunting climate crisis, we can begin to trace a discursive shift: metaphors of rescue, insurance, and medication are joined by notions of repair, restoration, and remediation.[3] Such language suggests the need to '[...] use humanity's extraordinary powers in service of creating a good Anthropocene'.[4] This, of course, begs the question of what this 'good Anthropocene' might look like and at whose service this approach of climate intervention might work.

In one version of the debate, climate engineering furnishes conservative programs of maintaining the economic-political status quo with a language of ecological sustainability, invoking 'green' ideals and ethical obligations. Initiatives, such as the *Ecomodernist Framework*, for example, envision scientific progress as a vital tool for positive control over the natural environment:

> [...] now, that we have the curse and blessing of knowing what's going on, unintentional is no longer an option. [...] We're left with intention, with conscious design, with engineering. We finesse climate or climate finesses us.[5]

Very much in line with this perspective, David Keith, one of the most prominently received voices on the topic, suggests reframing the debate over climate engineering:

> Geoengineering often seems a joyless choice between unpleasant alternatives. [...] I can't wholly embrace this view. It's an easy way out. About a million years after inventing stone cutting tools, ten thousand years after agriculture, and a century after the Wright Brothers flight, humanity's instinct for collaborative tool building has brought us the ability to manipulate our own genome and our planet's climate.[6]

This narrative of techno-scientific control, of course, has driven political interest in climate science for decades. And it has prepared the arrival of climate engineering in US climate policy since the turn of the new millennium, as we have seen in the previous chapter. Climate engineering appears in this context as an effort to make climate change 'legible', and therefore, 'amenable to containment in a way that preserves the current geo-political and economic order'.[7]

In a very different version of the debate, critical climate change scholarship and environmental activists have suggested the need to engage with climate engineering measures in a progressive way. Notions of repair, restoration, and remediation especially mobilise moral and ethical concerns, suggesting a human responsibility of restoring what has been disrupted.[8] Instead of cementing the economic and political status quo, climate engineering emerges in this context as a potential instrument for environmental and societal transformation. Climate engineering research has been promoted, for example, in the name of global social equity in tackling climate change. Govindasamy Bala and Aarti Gupta argue in this context that '[...] when our scientists tell us that the poorest people can be the greatest beneficiaries of solar geoengineering, we cannot dismiss them lightly'.[9] This issue of equity, in turn, has drawn attention to the lacking diversity and inclusiveness of the climate engineering research field and promoted calls for a further democratisation.[10] In particular, critical scholarship has suggested how marginal the position of countries from the global South remains in the current debate, emphasising the urgent need to diversify perspectives.[11]

CONCLUSION TO PART III

Finally, and in a particularly striking dynamic of the debate, climate engineering seems to have managed to mobilise a political constituency for climate science who officially rejects the very existence of anthropogenic climate change. A congressional hearing on climate engineering in November of 2017 documents Republican efforts to rid the concept of climate engineering from its reference point of climate change:

> I'd also like to take a moment to clarify any mischaracterizations about this hearing. The purpose of this hearing is to discuss the viability of geoengineering .... The hearing is not a platform to further the debate about climate change. We've had lots of that this session. Instead, its aim is to explore approaches and technologies that have been discussed in the scientific community and to assess the basic research needed to better understand the merits of these ideas. It is my hope that members will respect this focus so that we can have a meaningful discussion about geoengineering.[12]

The promise of control brings climate science back onto the Republican agenda here. In this version, climate engineering provides

> the potential to provide us with a whole new understanding and approach to atmospheric research. If we put aside the debates about climate change, we can support innovations in science that can create a better prospect for future generations.[13]

Arguably, this version of the debate is rather a return to 1950's perspectives than a historical synthesis of the role of climate science in the state. Yet, it demonstrates just how dynamic the political exploration of climate engineering remains. Techno-scientific intervention no longer seems to contradict programs of environmental safeguarding, and as a result, programs of environmental safeguarding no longer seem to contradict conservative political agendas of retaining the economic status quo. While all three of the above-mentioned versions of the climate engineering debate envision climate science as a tool for the state, this tool is fighting fundamentally different battles in each of them.

NOTES

1 See also Mitchell (2002).
2 See, e.g., Turner and Isenberg (2018). See also Nicholson and Thompson in Blackstock and Low (2019: 164f.).
3 See, e.g., Morrow (2014); Katz (2015); Baatz, Heyward, and Stelzer (2016); McLaren (2018a); King (2019); Fiekowsky (2019).
4 Asafu-Adjaye and others (2015).
5 Brand (2010: 19).
6 Keith (2000: 173).
7 Baker (2017: 20).
8 See, particularly, McLaren (2018a, 2018b).
9 Bala and Gupta (2017: 376).
10 See, e.g., Hulme (2014: 82).
11 Rahman and others (2018: 22–24); Biermann and Möller (2019). In 2020, the first African solar radiation management research papers were published by teams based at the University of Cape Town and Benin who were supported by SRMGI's DECIMALS Fund (Pinto and others (2020); Da-Allada and others (2020)). That same year, the first SRM modelling paper from the Middle East appeared (K. Karami and others (2020)).
12 Biggs in US House of Representatives, Committee on Science, Space, and Technology (2017: 4).
13 Chairman Weber in US House of Representatives, Committee on Science, Space, and Technology (2017).

CONCLUSION

SCIENTIFIC EXPERTISE AND THE POLITICS OF A 'BAD IDEA WHOSE TIME HAS COME'

CLIMATE ENGINEERING HAS GAINED POLITICAL TRACTION IN RECENT YEARS not by mobilising positive visions of techno-scientific innovation, but by promising to respond to a dire crisis – that is, by tackling increasingly dangerous climate change. The magnitude of this crisis seems to suggest that we are simply out of options when it comes to deciding how we do or do not wish to address it. Climate engineering emerges in this historical moment as something to try, perhaps crazy, perhaps impossible, but potentially, the 'least bad option we are going to have'.[1]

From this perspective, climate engineering seems to fit eerily well into the world that we live in today. It is a world that not only seems to be ridden with various globe-spanning problems, but that has also turned to scientific expertise for answers and solutions, for facts and fixes. As this book is being completed in the midst of the Covid-19 pandemic, notions of crisis, global problems, and 'grand societal challenges' have become increasingly central reference points for science-policy agendas.[2] Such notions seem to provide an ever more critical context for scientific expertise to prove and reaffirm its relevance to society as a whole. Expectations for science in society are perhaps higher than at any point in recent history.

This is something of an odd twist, given that we supposedly also live in times of 'post truth', conspiracies and denialism. In fact, it seems as if precisely such notions of 'post truth' have supercharged scientific facts with political expectations. Despite readily rehearsed assertions within the social sciences that we are aware of the societal embeddedness of scientific expertise and the limits of techno-scientific control, scientific facts and expertise have witnessed a momentous comeback in recent years. As a kind of external, a-political, and therefore, universal problem-solving authority, the seemingly definite assertions of science mobilise hopes of clarity and unity in such divided times.

Against this backdrop, the career of climate engineering provides a crucial source of insight, and an important call for caution at that. In this concluding chapter, I want to suggest how this book's analysis has not merely unpacked a highly controversial and somewhat curious debate in current climate policy contexts, but it has also shown how critically this debate speaks to the role of scientific expertise in contemporary politics.

(RE)ASSEMBLING THE CAREER OF CLIMATE ENGINEERING

At its core, this book has unpacked the rich history of a 'bad idea whose time has come'. Instead of essentialising climate engineering in its current form, the preceding chapters have sought to re-contextualise the making of this controversial governance object. The notion of the *career* of climate engineering has drawn attention to the historical contingency of what today is discussed as climate engineering. Retracing the career of climate engineering has helped to disentangle the various threads of scientific inquiry, national policy, and global geopolitical contexts that have, in hindsight, systematically brought us to the present point. It has enabled the turbulent trajectory of this 'bad idea' to be unpacked alongside its multiple temporalities, the diverse expert infrastructure that has assembled and stabilised it, and the epistemological categories which have defined climate engineering.

Before we dive into the broader significance of this analysis, let us briefly revisit some of its core findings. The book has revealed at least two defining historical threads (or even temporalities) in the career of climate engineering. Fig. 8.1 suggests how these two defining temporalities – suggested via the brackets – can be further differentiated into at least four distinct historical settings which define the career of climate engineering.

CONCLUSION

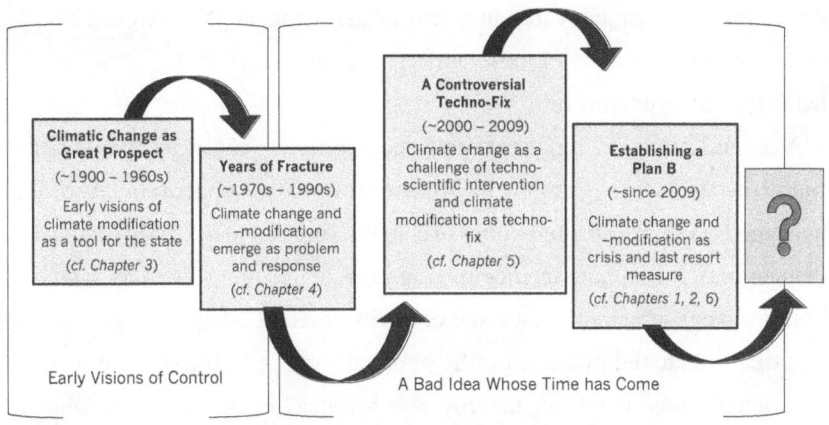

FIG. 8.1 The Career of Climate Engineering in US Policy

The first historical thread of this career of climate engineering is rooted in the history of climatology. In Chapter 2, we have seen how notions of targeted climate intervention have their historical origins in isolated projects of scientific curiosity at the turn of the twentieth century. Historical scholarship has illustrated in this context that before initial findings on the possibility of human impacts on climatic change were problematised, they provoked positive techno-scientific visions of the targeted modification and control of the climate. Climate change thus emerged as a 'great prospect'. This dynamic then came into full swing with significant observational and modelling progress during the second half of the twentieth century.

During this time, a massive infrastructure of organisations, professionals, programs, and technical equipment stabilised the 'quest for perfectly accurate machine forecasts and [...] perfectly accurate data acquisition'.[3] This infrastructure transformed the atmosphere into a subject that could not only be qualitatively described and mapped, but also 'rendered calculable', and therefore – this was the hope – controllable.[4] This infrastructure provided the critical observational basis for assembling climatic change as a governance object that would lend itself to deliberate techno-scientific modification. Around the 1960s, we can observe an incremental shift in how visions of climate modification were problematised, when the distinction of deliberate and inadvertent climate modification appeared in scientific assessments. What would later be posited as problem and response, emerged as two sides of the same coin: as challenge and opportunity.

A second historical thread of the career of climate engineering was added around the 1970s through to the 1990s when the politicisation of climate change challenged carefully nurtured hopes of climate control and fractured established alliances between climate science and the state. In Chapter 4, we have seen how climate change emerged as an issue of environmental safeguarding; it was an issue that threatened to question the political and economic status quo. Only incrementally was climate engineering able to regain political traction. Alliances between climate science and the state now evolved around coming to grips with defining and understanding the problem of climate change. Climate engineering in this context appeared as a potential response measure to an increasingly urgent problem of global political significance. Finally, at the dawn of the new millennium, climate engineering moved from the margins of natural scientific assessments of climate change to the heart of fiercely contested congressional debates over how to tackle it best. By the end of its first decade, the concept had arrived in US climate policy as a challenge in its own right: climate engineering now took shape as the least evil in a hopeless situation.

This *career* of climate engineering thus corresponds to historically shifting modes of making sense of and problematising climatic change in relatively distinct historical settings. The boxes in Fig. 8.1 provide a strongly simplified overview. These historically particular modes of making sense of and problematising climatic change, also correspond to shifting configurations or alliances between climate science and the state. These alliances were forged by military and geopolitical challenges during the first half of the twentieth century by environmental challenges and the politicisation of climate change during the second half of the twentieth century, and by the notion of a climate emergency since the dawn of the new millennium. We have seen how the making of this 'bad idea' has mutually linked science and politics; how historically contingent visions of what is today discussed as climate engineering have provided a continuous, yet shape-shifting node in linking climate science and the state across different contexts and throughout different times; how climate engineering, in other words, can be understood as the result of highly specialised scientific and political processes, *as well as* their growing interdependence.

This outlook then questions some of the prominent dichotomies that have come to define the current debates over climate engineering. Particularly, it questions the narrative of the historical fracture that the recent rise of climate engineering has implied for existing climate policy agendas and established alliances between climate science and the state. A bold reading of this book's account might even suggest that we tell this story the other way around, namely, as a story of how visions of climate engineering turned into explorations of anthropogenic climate change. Such a bold reading would suggest that it was not so much climate engineering that disrupted established climate change agendas, but rather that it was climate change that disrupted established climate engineering agendas and that it was the politicisation of anthropogenic climate change that fundamentally questioned established science-politics alliances around hopes of techno-scientific control.

More modestly, however, this book has shown that climate engineering has not emerged out of thin air. The book has embedded climate engineering in the larger history of efforts to understand and govern climatic change. Through the lens of this career of climate engineering, we get to see just how closely entwined these stories of efforts to understand and efforts to deliberately modify (even control) climatic change have been. This means that, as a response, climate engineering rests on a particular way of defining the issue at hand. This 'bad idea' grew out of a distinct mode of assembling climate change. Problem and response, in other words, are interrelated.

This does not, of course, delegitimise a critique of climate engineering, let alone render these measures unequivocally desirable. On the contrary, the analysis suggests that a meaningful engagement with (and critique of) climate engineering, must go further. It suggests that instead of singling out and essentialising climate engineering as the somewhat crazy, yet inevitable last resort measure, we must explore this interrelation between assembling and addressing, between understanding and governing climate change much more thoroughly. A meaningful engagement with climate engineering as a controversial policy measure essentially hinges on a thorough assessment of the modes of expert observations which have assembled it and the expert infrastructure which have stabilised this particular problem formulation. A meaningful critique of climate engineering cannot disregard the problematisation of climate change which these measures rest on and respond to. This is true even more so if we do in fact consider climate engineering as a 'bad idea' or as 'barking mad'.[5] The previous chapters suggest that it is high time to move beyond distinctions of good and bad when it comes to climate science and climate engineering and to instead unpack their joint histories and infrastructures. I want to make two points on the matter.

ON PLURALISING EXPERT OBSERVATIONS

To begin with, the perspective on the career of climate engineering advanced in this book suggests why and how it might be productive to pluralise policy-relevant expert perspectives on climate change. It suggests the merit of diversifying expert modes of observing and assembling the issue of climatic change. By making this suggestion, the analysis speaks to a body of literature which has pointed out that more climate science does not in fact lead to more effective policy action. Despite the vast amounts of resources that climate policy programs have poured into understanding the mechanical, physical, and chemical grounds of anthropogenic climate change, there has been little progress in addressing the negative impacts of climate change for society and the environment.[6] Recent accounts in environmental history go even further, suggesting that progress in our climatological understanding of global warming not only failed to spur mitigation of the issue, but – quite to the contrary – directly corresponded to a spectacular acceleration of the crisis. There is now more CO_2 being emitted to the atmosphere on a daily basis than ever before in human history.[7]

How might we make sense of this? Why has this ever more differentiated picture of the climatological intricacies of global warming had so little effect on halting its causes? An obvious response to these questions is that, quite simply, understanding is not the same as acting, that knowing about an issue is not the same as doing something about it, and that, by extension, scientific facts and findings do not affect political action in any linear or targeted sense. Following this line of reasoning, science policy literature has diagnosed an oversupply of useless expertise, pointing to the irrelevance of the provided expertise to effectuate policy change.[8] The authors find that a general mismatch between the provided expertise and the needs of political decision-makers provokes the systematic failure to generate 'relevant and usable scientific information' in this case of climate change.[9] The scientists, put differently, follow a different logic in their work than the political decision-makers, and therefore, supply and demand of expertise are not balanced and need to be reconciled to ensure a better societal outcome to act on this issue.

I want to suggest that this book's analysis of the career of climate engineering complements this diagnosis. Specifically, it aids in further differentiating the notion of what we consider as politically 'relevant' scientific expertise by emphasising the interrelation between science and politics in assembling and addressing societal problems.

Despite being framed as a 'Plan B', climate engineering evolved from persistent political efforts to cultivate and harness climatological expertise as a means to make climatic change politically legible and governable. Somewhat paradoxically, the rise of this 'bad idea' suggests how climatological modes of assembling the issue have proved politically effective as much as misleading in the sense that they suggested the possibility of their political control. We have seen how, over the years, expert committees and agencies have assembled the issue of anthropogenic climate change as an issue that concerns the understanding and governance of an exceedingly complex 'climate system', compounded of not only the atmosphere, but also the ocean, large ice-shields, such as sea ice or glaciers, the pedosphere, as well as the marine and terrestrial biospheres and many more parameters.[10] The career of climate engineering suggests how this particular mode of observing climate change not only turned the issue into a challenge to utilise ever more precise scientific observation, measurement and prediction of a complex climatological system. But also, it demonstrates how this mode of observing the climate corresponded to and even directly fuelled hopes of being able to deliberately modify and control the parameters of this very system. Climate engineering, in this sense, emerged as a project of climatological cultivation and control. This case of climate engineering thus suggests that the issue of scientific expertise in climate policy is not so much useless, but, rather to the contrary: it involves a somewhat distorted overreliance on one particular expert mode of observation on the issue at hand. To put it bluntly, the career of climate engineering suggests just how consequential the political cultivation of climatological expertise has been.

This, of course, is not to say that the political consideration and support of climatological expertise, let alone the mere progress of climatology, single-mindedly led us to the prospect of climate engineering. However, it is to stress the complex interplay between science and politics in assembling and addressing societal problems. It is to emphasise that the history and particular trajectory of climate engineering is intimately linked to the history and particular trajectory of alliances between climate science and the state. Science and politics do not merely meet in this context as evidence base on the one hand, and decision-making authority on the other. Both are coupled at the upstream; that means, they are mutually constitutive in assembling societal problems and devising respective response measures. The question then is not merely one of more or less useful facts, but a question of what kinds of facts for what kinds of politics, and, vice versa, what kind of politics for what kind of facts.[11]

The turbulent trajectory of climate engineering reminds us that it matters how we choose to look at the issues of our time. It reinforces to us that we are precisely not out of options. If anything, this book's account suggests we need to consider alternative, additional, and more diverse perspectives in making sense of and addressing the issue of climate change. While climate science is obviously essential to this endeavour, it can only solve part of the puzzle. To change course, it seems essential to broaden disciplinary vistas and avoid a kind of tunnel vision onto the climate 'out there'. In this context, scholars from various disciplinary backgrounds have repeatedly emphasised the continued lack of social scientific perspectives within climate policy.

In the early 1990s, environmental scholars Peter Taylor and Frederick H. Buttel already began to ask, *How do We Know We Have Global Environmental Problems?* In this essay, the authors contrast the problem-defining authority of climatological expertise with the social precondition of the climate change challenge:

Of course, global change researchers know that climate change is a social problem, since it is through industrial production, transport and electrical generation systems and tropical deforestation that societies generate greenhouse gases. Nonetheless, it is physical change – the mechanical and inexorable greenhouse effect – that is invoked to promote policy responses and social change.[12]

This important observation continues to resound in more recent accounts of the potential role and importance of the social sciences and humanities in crafting meaningful climate change policy.[13] Such accounts emphasise that climate change is not merely a biophysical process. They illustrate how climate change is an issue of social and power relations engraved in human infrastructures; it is an issue of how we live and sustain ourselves, how we work and commute, and it is an issue of international relations, economic systems, and political order. In his comprehensive account on *The Rise of Steam Power and the Roots of Global Warming*, Andreas Malm, for example, explores the critical role of labour relations in driving the issue of climate change. He writes that

> Anthropogenic climate change – this is part of its very definition – has its roots *outside* the realm of temperature and precipitation, turtles and polar bears, inside a sphere of human praxis that could be summed up in one word as *labour*.[14]

The task, then, is not to explore the ways in which climate and climate change have defined history, but to explore how history has shaped climate: 'in a warming world, causation runs, at least initially, from company to cloud'.[15] To change course, in other words, it is essential to understand these kinds of drivers of climatic change, to explore more thoroughly how we ended up in this mess in the first place.

The funders of scientific research have also recognised the lack of social scientific perspectives on the problem of climate change. In its 2012 review of the government's new ten-year strategic plan for the federal development of climate change expertise, the US National Research Council emphasised that the plan's 'proposed broadening [...] to better integrate the social and ecological sciences' would be 'essential for [...] *understanding* and *responding* to global change'.[16] It openly criticised that the responsible agencies 'have insufficient expertise in these domains and lack clear mandates to develop the needed science'.[17]

Almost a decade later, however, the issue seems to persist. Pluralising expert modes of observing and assembling climate change as a societal problem would thus necessitate a more consistent and radical mode of interdisciplinarity. It cannot merely imply charging the social sciences and humanities with illuminating the consequences of techno-scientific interventions. Chapter 6 has suggested that the situation is much more complicated than that. A change of perspective would require inviting competing – even conflicting – modes of observation. It would imply strengthening diverging modes of making sense of societal issues. It would require not only hunting for 'facts' but allowing for ambiguity and complexity.

FROM SCIENTIFIC CONSPIRACIES, CLIQUES, AND MAFIAS TO MATCHED STRUGGLES

The perspective on the career of climate engineering provided by this book suggests a second area where critical attention is required to move beyond simplistic distinctions between good and bad when it comes to climate science and climate engineering – accounts that reduce climate engineering to the concerted agenda of individual scientists. The preceding chapters have illustrated that neither the discovery of global warming as a societal problem of global political significance, nor the rise of climate engineering as a controversial 'risk management strategy'[18] for tackling this issue are merely the result of a scientific conspiracy, sinister or otherwise. Visions to technically modify, deliberately alter, or engineer the climate have not been forced onto the political agenda in an orchestrated plot by a group of scientific experts. Rather, the previous chapters have suggested that scientific expertise has to be understood as relational. Who gets to 'speak' to politics, who appears as a scientific spokesperson, which experts and what kind of expertise are heard by policymakers – and thus define political agendas – depends just as much on political selection processes as on the experts themselves. To be politically relevant, that is, to bear in one way or another on political processes, scientific expertise must resonate in and be internalised by the political system.

Rather than overemphasising the relevance of individual scientists and their personal agenda in pushing policy programs and forcing their perspectives onto the political realm, the book has drawn our attention to the settings which have channelled these experts onto the political stage. It has shed light on the *expert infrastructure*, that is, the channels and pathways, arenas and corridors, effectively structuring alliances between climate science and the state, and thus giving these alliances their particular form and shape. Therefore, this perspective on the career of climate engineering not only helps make sense of the recent rise of this 'bad idea whose time has come' by opening our gaze onto this multi-layered expert infrastructure, but the analysis also speaks more broadly to science policy studies and the study of expertise in politics.

By following the career of climate engineering as it has defined science-state alliances in different contexts and at different times, we gain a comparative perspective on the various settings defining these alliances and linking scientific expertise to politics. The general insight from mapping this climate engineering expert infrastructure is that the political relevance of scientific expertise can be differentiated along the particular settings which connect it to the policy process. Put differently, the various settings which channel scientific expertise to politics correspond to respectively distinct roles of scientific expertise in shaping the politics of climate engineering. There are congressional expert witnesses serving as 'masked agenda setters', pushing a controversial measure seemingly incidentally onto the congressional agenda before policymakers take an official stance on the issue (see Chapter 5). There are also various settings in which policymakers invite or commission different forms of 'staged advice', defining the issue at stake as a 'matter of facts' (see Chapters 2, 4 and 6), and there are expert organisations and programs within the federal bureaucracy that institutionalise the political cultivation of issue-relevant expertise within the political realm (see Chapters 2 and 6).

Two such expert settings, in particular, appear to be important components of the climate engineering expert infrastructure. The first expert setting concerns what I suggest we call arenas of 'staged advice'. We have seen how these arenas provide the relevant context for linking scientific expertise from beyond the federal infrastructure to the political realm. This is an expert setting which indeed made scientists into a kind of spokespeople for assembling policy issues. Both the politicisation of climate change during the 1980s, as well as the politicisation of climate engineering during the early 2000s, became visible to the public primarily by means of a highly visible group of experts who were 'raising the alarm'[19] and defining the relevant 'facts' about these issues respectively. The fact that the literature has somewhat ironically described these experts as conspiracies, cliques, or mafias is instructive in this context but not for blaming or praising the political stance of these individual scientists and experts. Instead, these metaphors are telling as they suggest an underlying logic, an implicit script or system that guides the concerted agenda of these scientific experts and explains their visibility and impact in shaping the political agenda.

Chapters 2 and 6 speak to this assumption by unpacking just how carefully these arenas of staged advice are orchestrated. The chapters suggest that the political system has developed two forms of such staged advice that serve to determine politically relevant experts and channel scientific spokespeople onto the political stage, namely scientific assessments and legislative inquiries. In the context of climate change, as well as climate engineering, scientific experts shaped the political agenda on these issues by testifying before Congress and contributing to scientific assessment reports. Both forms of staged advice institutionalise the selection of politically relevant scientific expertise differently. Both embody different modes of linking scientific to political observations. Scientific assessments seek to harvest scientific excellence for policy questions. While policymakers may ask the questions (and commission or approve the assessments), the assessment bodies strive to institutionalise criteria of scientific quality in the selection of viable experts and the production of these assessments. Scientific assessments and their respective lists of experts thus document how the politicisation of visions to modify and intervene in the global climate have historically connected to inner-scientific structures of disciplinary distinctions, research schools and programs, methodological and conceptual outlooks.

Legislative inquiries, in contrast, give much more leeway to the policymakers and committees in selecting who qualifies as a relevant expert. This selection process is precisely not specialised on the selection of *scientific* experts.[20] It follows additional criteria other than scientific output and academic reputation, which makes the prominent status of scientists (even academic scientists) as spokespeople for the politicisation of climate change and climate engineering all the more noteworthy. What is more, in the politicisation of climate change as well as climate engineering, the experts testifying before Congress were often also quite visible and outspoken on these issues in the media. The exact connection between these two contexts of appearance would have to be studied systematically in future research. Does political expert status explain media visibility or does media visibility explain political expert status? Aside from scientific and technical credibility and media appearances, the literature points to a plurality of selection criteria that are at play in this context.[21]

The second critical component of the climate engineering expert infrastructure is an expert setting that I suggested we call 'science for national needs'. We have seen how, building on the invitation of staged advice, the political system internalises scientific expert modes of observation into the federal infrastructure. The state thus stabilises and institutionalises a particular gaze on the issue at hand, seeking to actively steer or cultivate politically relevant expertise within its own infrastructure. This expert setting thus does not channel external expertise towards the political system, but instead cultivates relevant expertise within its own bounds.

Through the lens of this expert infrastructure, climate engineering hardly appears as something controversial and new. From this angle, the rise of this 'bad idea' instead emerges as rooted in the organisations and programs that have defined the US political problematisation of climate change for decades. This includes agencies such as the IPCC, central agencies of the US Global Change Research Program (USGCRP), such as NASA, NOAA, EPA, DOE, as well as WHOI or LLNL. Many of these organisations have been established even before anthropogenic climate change had even been politicised as an issue in its own right. They essentially represent the institutionalisation of atmospheric and oceanographic research within the federal bureaucracy.

OUTLOOK: FROM FOLLOWING THE ACTORS TO FOLLOWING THE PROBLEMS?

This book presented an analysis of the role of scientific experts and expertise in politics by approaching the matter somewhat inversely. Instead of merely following the experts, it placed an object of expert work, namely climate engineering, front and centre. Connecting to the basic tenet of 'following the actors', it sought to give climate engineering a life of its own.[22] The book followed climate engineering through shifting historical contexts, through diverse expert settings, and different expert modes of observation to get a sense of how it became what it is today. In other words, it sought to understand how it became assembled as a controversial policy tool to fight the problem of anthropogenic climate change.

As global problems and societal challenges become increasingly important reference points in determining the societal status of science and the political relevance of scientific expertise, this approach promises to provide a differentiated picture of the complex interplay of science and politics as two distinct, yet increasingly interdependent, realms of society. I want to end by reflecting on the potential merit of this approach for making sense of the status of science in society – not only as an academic enterprise within science studies, but more generally, to suggest how it might help us grapple with the current situation we find ourselves in, which is beset by so many often overlapping global challenges, as laid out at the beginning of this chapter.

Approaching the science-politics interrelation via the historical trajectories or 'careers' of societal problems sheds light on the contingent and reflexive nature of this interplay. The analysis in this book demonstrates that climate engineering is neither the result of a simple steering (politics → science), nor the result of an informing or advisory relationship (science → politics). Instead, it emerges precisely from the mutual observation of both societal realms. This approach, then, suggests how the interrelation between science and politics amounts to more than the sum of its parts. Giving climate engineering a life of its own in this sense means considering the inherently dynamic trajectory that this object of expertise unfolded.

CONCLUSION

We saw how this reflexive relation is written into a diverse infrastructure of expert settings and various kinds of expertise that emerge at the interface of science and politics, and that have given climate engineering its particular shape over the years. Following the career of climate engineering in this sense sheds light on how settings of expert policy advice complement networks of epistemic communities or independent scientific agencies in linking science and politics. At the same time, it embeds this expert infrastructure and these forms of expertise in their particular historical context. The witness lists and expert assessments, and the programs and missions of expert agencies, for example, document how the political problematisation of climatic change and intervention has shifted over the years. This approach thus combines an interest in overarching structures that define the interrelation of science and politics across time and space with an interest in the historical genesis and evolution of these very structures. Finally, following this career of climate engineering suggests how, vice versa, this object of expertise bears consequences for science and politics respectively. This is a theme which has been less thoroughly explored in this book and needs to be subject to future research. The book can only hint at how climate engineering fosters changes in inner-scientific structures and political landscapes, how different visions of modifying and intervening in the global climate generate new publication networks, research communities, and gradually also formal scientific programs, how it gives rise to new political constituencies regarding climate change, and how it formulates new political categories.

Therefore, the analytical approach of this book might also help us make sense of the status of science in society today. It may help us grapple with the crisis narrative that is so often attached to major contemporary global challenges – to return to the themes with which this chapter and the book as a whole began. To put it bluntly, the story that this book tells suggests that we are precisely not out of options. Instead, the perspective that this analysis has opened up encourages a more productive way of engaging with science in society. It suggests a more productive vision for the role and status of science in addressing the issues of our time.

In shedding light on the career of climate engineering, the book's analysis suggests that advancing scientific expertise in the name of emergency or crisis is unproductive if it serves to suggest a lack of agency in the face of the issues that societies face today. It is unproductive, in other words, if it is mobilised as a means of closing down democratic dispute or options of engagement and political controversy. We have seen that scientific expertise is not external to the issues that it promises to tackle. Instead, the process of making sense of societal problems and devising response measures, of assembling and addressing governance objects, is reflexively linked. Scientific expertise might seem difficult to argue with. This book suggests that it shouldn't be. Mobilising scientific expertise in the name of tackling crises and societal issues can be productive precisely if it serves to foster reflexivity and a change of direction. Instead of emphasising factual constraints and closing down possible futures, it should be mobilised to diversify them.

Promoting climate engineering as a necessary evil, then, is not only problematic because it proposes a narrative of scientific control in the face of the dangers of climate change. It is also actively misleading if it serves to deflect attention away from an understanding of how we got here, via a rich and multi-layered history, involving the cultivated structures that have systematically brought forth this 'bad idea whose time has come'. Taking this rich history of climate engineering seriously is also essential for challenging the narrative of a future without choices. As I suggested in the introduction, as much as politics might sometimes allude to external urgencies that force our hands, climate engineering is not infused into the political process by the external urgency of dangerous climate change. It arrived here from within: this particular vision of making sense of and responding to climatic change has a historical legacy and system.

NOTES

1 Specter (2012).
2 See, e.g., Varmus and others (2003); Brooks and others (2009); Calvert (2013); Hicks (2016); Kaldewey (2018).
3 Fleming (2010: 192).
4 Baker (2017: 11).
5 Pierrehumbert (2015).
6 See, e.g., Sarewitz and Pielke (2007: 11–12).

7 Locher and Fressoz (2012); Bonneuil and Fressoz (2016); Malm (2016).
8 See, e.g., Sarewitz (2004); Sarewitz and Pielke (2007).
9 Sarewitz and Pielke (2007: 14).
10 Gramelsberger and Feichter (2011: 11–15); see also Hulme (2014).
11 See also Hulme (2012).
12 Taylor and Buttel (1992: 410).
13 To name but a few contributions in this context, see, e.g., Adger and others, (2013); Hulme (2014); Weaver and others (2014); Hackmann, Moser, and St. Clair (2014); Victor (2015); Malm (2016).
14 Malm (2016: 6, emphasis in original).
15 Malm (2016: 6, emphasis in original).
16 US National Research Council (2012: 1–2, emphasis added).
17 US National Research Council (2012: 1–2, emphasis added).
18 US Government Accountability Office (2010b: 2). See also US Government Accountability Office (2011).
19 Turner and Isenberg (2018: 33).
20 Although the literature points to several reasons why the selection of scientists and technical experts makes political sense (see, e.g., Smallman 2020).
21 See, e.g., Keller (2009); Boswell (2012).
22 For the notion of following the actors, see, e.g., Latour and Woolgar (1979); Pinch and Bijker (1984); Latour (1987); Latour (1999).

APPENDIX
DOCUMENT CORPUS

	YEAR	TITLE	AUTHOR	DOC TYPE	CLIMATE ENGINEERING CONTEXT
1	1990	Changing Climate and the Coast, Volume 2	NOAA	Report	Umbrella term
2	1997	Global Climate Change	Senate Committee on Environment and Public Works	Hearing	Umbrella term
3	2003	What are the Administration's Priorities for Climate Change Technology?	House Committee on Science	Hearing	CDR
4	2004	Federal Register Volume 69, Number 77	Department of Energy	Notice	CDR
5	2006	Department of Energy's Plan for Climate Change Technology Program	House Committee on Science	Hearing	Umbrella term, CDR
6	2006	Climate Change Technology Research	House Committee on Government Reform	Hearing	Umbrella term
7	2007	The Future of Coal	Senate Committee on Energy and Natural Resources	Hearing	Umbrella term, CDR
8	2007	Effects of Climate Change and Ocean Acidification on Living Marine Systems	Senate Committee on Commerce, Science, and Transportation	Hearing	CDR
9	2007	Voluntary Carbon Offsets: Getting What You Pay For	House Sel. Committee on Energy Independence and Global Warming	Hearing	CDR
10	2007	An Examination of the Impacts of Global Warming on the Chesapeake Bay	Senate Committee on Environment and Public Works	Hearing	Umbrella term
11	2008	United States-China Relations in the Era of Globalization	Senate Committee on Foreign Relations	Hearing	Umbrella term
12	2008	Report on Global Change Research Improvement Act of 2007	Senate Committee on Commerce, Science, and Transportation	Report	CDR

	YEAR	TITLE	AUTHOR	DOC TYPE	CLIMATE ENGINEERING CONTEXT
13	2009	Commerce, Justice, Science, and Related Agencies Appropriations for 2010	House Committee on Appropriations	Appropriations Hearing	Umbrella term, CDR
14	2009	The Role of Offsets in Climate Legislation	House Committee on Energy and Commerce	Hearing	CDR
15	2009	Innovative, Non-Geologic Applications for the Reuse of Carbon Dioxide	Senate Committee on Appropriations	Appropriations Hearing	CDR
16	2009	Commerce, Justice, Science, and Related Agencies Appropriations Bill for 2010	House Committee on Appropriations	Report	CDR
17	2009	Public Transportation: A Core Climate Solution	Senate Committee on Banking, Housing and Urban Affairs	Hearing	Umbrella term
18	2009	American Clean Energy and Security Act of 2009 (H.R.2454)	Henry Waxman (Sponsor), Markey Edward (Sponsor)	Proposed Bill (Introduced in Senate (IS))	CDR
19	2009	Climate Services: Solutions from Commerce to Communities	Senate Committee on Commerce, Science and Transportation	Hearing	Umbrella term
20	2009	Drought, Flooding and Refugees	Senate Committee on Foreign Relations	Hearing	Umbrella term
21	2009	Building US Resilience to Global Warming Impacts	House Select Committee on Energy Independence and Global Warming	Hearing	Umbrella term
22	2009	Geoengineering Parts I, II, and III	House Committee on Science and Technology	Hearing	Umbrella term, CDR, SRM
23	2009	Carbon Dioxide Capture Technology Act of 2009 (S.2744)	John Barrasso (Sponsor)	Proposed Legislation (Introduced in Senate (IS))	CDR
24	2009	Policy Options for Reducing Greenhouse Gas Emissions	Senate Committee on Energy and Natural Resources	Hearing	CDR
25	2010	Legislative Branch Appropriations for 2011	House Committee on Appropriations	Appropriations Hearing	Umbrella term
26	2010	Fiscal Year 2011 R&D Budget Proposals (EPA & NOAA)	House Committee on Science and Technology	Hearing	CDR
27	2010	Combating Climate Change in Africa	House Committee on Foreign Affairs	Hearing	Umbrella term
28	2010	Congressional Record Volume 156, Number 71: America Competes Reauthorization Act of 2010 (H.R. 5116)	House of Representatives	Congressional Record	CDR

	YEAR	TITLE	AUTHOR	DOC TYPE	CLIMATE ENGINEERING CONTEXT
29	2010	Engineering the Climate	House Committee on Science and Technology	Report	Umbrella term, CDR, SRM
30	2010	America's Energy Security, Jobs and Climate Challenges	House Sel. Committee on Energy Independence and Global Warming	Hearing	Umbrella term
31	2010	Federal Register Volume 75, Number 237	Environmental Protection Agency	Final Rule	Umbrella term, CDR
32	2011	Carbon Dioxide Capture Technology Prize Act of 2011 (S.757)	Jeff Bingaman (Sponsor)	Proposed Legislation	CDR
33	2011	Carbon Capture and Sequestration Legislation	Senate Committee on Energy and Natural Resources	Hearing	CDR
34	2011	Domestic Policy Implications of the UN Declaration on the Right of Indigenous Peoples	Senate Committee on Indian Affairs	Hearing	Umbrella term
35	2011	Senate Report 112-33: Carbon Dioxide Capture (to accompany 757)	Senate Committee on Energy and Natural Resources	Report	CDR
36	2011	House Report 112-169	House Committee on Appropriations	Report	CDR
37	2011	Need for Continued Innovation in Forecasting and Prediction	Senate Comitee on Commerce, Science, and Transportation	Hearing	Umbrella term
38	2012	Congressional Record Volume 158, Issue 128: Climate Change	Sheldon Whitehouse, United States Senate	Congressional Record	Umbrella term
39	2013	Climate Change: It's Happening Now	Senate Committee on Environment and Public Works	Hearing	Umbrella term
40	2014	Department of Energy: Science and Technology Priorities	House Committee on Science, Space and Technology	Hearing	Umbrella term
41	2015	Examining EPA's Proposed Carbon Dioxide Emissions Rules	Senate Committee on Environment and Public Works	Oversight Hearing	Umbrella term
42	2015	Energy and Water Development Appropriations Bill 2016 (Senate Report)	Senate Committee on Appropriations	Report	CDR
43	2015	Federal Register Volume 80, Issue 205	Environmental Protection Agency	Final Rule	CDR
44	2015	Federal Register Volume 80, Issue 205	Environmental Protection Agency	Proposed Rule	CDR
45	2015	Congressional Record Volume 161, Issue 173	United States Senate	Congressional Record	Umbrella term

	YEAR	TITLE	AUTHOR	DOC TYPE	CLIMATE ENGINEERING CONTEXT
46	2015	Congressional Record Volume 161, Issue 184: Fossil Energy Research and Development	United States Senate	Congressional Record	CDR
47	2016	Congressional Record Volume 162, Number 112: Climate Change	Sheldon Whitehouse, United States Senate	Congressional Record	Umbrella term
48	2017	Developing and Deploying Advanced Clean Energy Technologies	Subcommittee on Clean Air and Nuclear Safety of the Committee on Environment And Public Works United States Senate	Hearing	CDR
49	2017	Walker and Winberg Nominations	Committee on Energy and Natural Resources United States Senate	Hearing	CDR
50	2017	Geoengineering: Innovation, Research, and Technology	Committee on Science, Space, and Technology, House of Representatives	Hearing	Umbrella term, CDR, SRM
51	2017	Congressional Record Volume 163, Number 200: Introduction of the Geoengineering Research Evaluation Act Of 2017	House of Representatives (Jerry McNerney)	Congressional Record	Umbrella term
52	2017	The Geoengineering Research Evaluation Act of 2017	Jerry McNerney (sponsor)	Proposed Bill (Introduced in House (IH))	Umbrella term; CDR, SRM
53	2018	Congressional Record Volume 164, Number 1: Climate Change	Sheldon Whitehouse, United States Senate	Congressional Record	CDR
54	2018	The Utilizing Significant Emissions with Innovative Technologies Act (USE IT Act) (S. 2602)	John Barrasso, Sheldon Whitehouse, Shelley Capito, Heidi Heitkamp (sponsors)	Proposed Bill (Introduced in Senate (IS))	CDR
55	2018	Business Meeting: S. Hrg. 115–578	Committee on Environment and Public Works, United States Senate	Business Meeting	CDR
56	2018	Legislative Hearing on S. 2602, The Utilizing Significant Emissions with Innovative Technologies Act, or USE IT Act	Committee on Environment and Public Works, United States Senate	Hearing	CDR
57	2018	Using Technology to Address Climate Change	Committee on Science, Space, and Technology, House of Representatives	Hearing	Umbrella term

	YEAR	TITLE	AUTHOR	DOC TYPE	CLIMATE ENGINEERING CONTEXT
58	2018	Fossil Energy Research and Development Act of 2018 (H. R. 5745)	Marc Veasey, David McKinley, Eddie Bernice Johnson (Sponsors)	Proposed Bill	CDR
59	2018	The Modernizing America with Rebuilding to Kickstart the Economy of the Twenty-first Century with a Historic Infrastructure-Centered Expansion Act (MARKET CHOICE Act) (H. R. 6463)	Carlos Curbelo, Brian Fitzpatrick (Sponsors)	Proposed Bill (Introduced in House (IH))	CDR
60	2018	Utilizing Significant Emissions with Innovative Technologies Act, Or USE IT Act (to accompany S. 2602)	John Barrasso	Report	CDR
61	2019	Healthy Oceans and Healthy Economies: The State of our Oceans in the 21st Century	Committee on Natural Resources, House of Representatives	Oversight Hearing	Umbrella term, CDR
62	2019	The State of Climate Science and Why It Matters	Committee on Science, Space, and Technology, House of Representatives	Hearing	Umbrella term
63	2019	Hearing to Examine S. 383, the Utilizing Significant Emissions with Innovative Technologies Act, and the State of Current Technologies that Reduce, Capture, and Use Carbon Dioxide	Committee on Environment and Public Works, United States Senate	Hearing	CDR
64	2019	The Status and Outlook of Energy Innovation in the United States	Committee on Energy and Natural Resources, United States Senate	Hearing	CDR
65	2019	Business Meeting: S. Hrg. 116-17	Committee on Environment and Public Works, United States Senate	Business Meeting	CDR
66	2019	Time for Action: Addressing the Environmental and Economic Effects of Climate Change	Committee on Energy and Commerce, House of Representatives	Hearing	Umbrella term
67	2019	Sea Change: Impacts of Climate Change on our Oceans and Coasts	Committee on Science, Space, and Technology, House of Representatives	Hearing	CDR
68	2019	Climate Change National Security Strategy Act of 2019 (H. R. 1201)	Stephen Lynch and others (Sponsor)	Proposed Bill (Introduced in House (IH))	Umbrella term

	YEAR	TITLE	AUTHOR	DOC TYPE	CLIMATE ENGINEERING CONTEXT
69	2019	Congressional Record Volume 165, Number 45: The Green New Deal	John Barrasso	Congressional Record	CDR
70	2019	Hearing to Examine S.747, Diesel Emissions Reduction Act of 2019	Committee on Environment and Public Works, United States Senate	Hearing	CDR
71	2019	Enhancing Fossil Fuel Energy Carbon Technology Act of 2019 (S. 1201)	United States Senate	Proposed Bill (Introduced in Senate (IS))	CDR
72	2019	Business Meeting: S. Hrg. 116-18	Committee on Environment and Public Works, United States Senate	Business Meeting	CDR
73	2019	Utilizing Significant Emissions with Innovative Technologies Act (to accompany S. 383)	John Barrasso	Report	CDR
74	2019	Congressional Record Volume 165, Number 99: Carbon Capture Prize Act	Grace Meng	Congressional Record	CDR
75	2019	Fossil Energy Research: Enabling Our Clean Energy Future	Committee on Science, Space, and Technology, House of Representatives	Hearing	CDR
76	2019	Climate Action Rebate Act of 2019 (H. R. 4051)	James Panetta, Thomas Suozzi (sponsors)	Proposed Legislation (Introduced in House (IH))	CDR
77	2019	Congressional Record Volume 165, Number 120: Climate Change	United States Senate: Sheldon Whitehouse	Congressional Record	CDR
78	2019	Earth's Thermometers: Glacial and Ice Sheet Melt in a Changing Climate	Committee on Science, Space, and Technology, House of Representatives	Hearing	Umbrella term
79	2019	Business Meeting (Senate Hearing 116-138)	Committee on Environment and Public Works, United States Senate	Business Meeting	CDR
80	2019	Congressional Record Volume 165, Number 115: National Defense Authorization Act for Fiscal Year 2020	House of Representatives	Congressional Record	CDR
81	2019	Congressional Record Volume 165, Number 124: Support Increased Domestic Energy Production	House of Representatives	Congressional Record	CDR

	YEAR	TITLE	AUTHOR	DOC TYPE	CLIMATE ENGINEERING CONTEXT
82	2019	Congressional Record Volume 165, Number 125: Recognizing Sheldon Whitehouse's 250th Climate Change Speech	United States Senate	Congressional Record	CDR
83	2019	Energy and Water Development Appropriations Bill, 2020 (to accompany S. 2470)	Lamar Alexander (Senate Committee on Appropriations)	Report	CDR
84	2019	Solving the Climate Crisis: Reducing Industrial Emissions Through US Innovation	Select Committee on the Climate Crisis, House of Representatives	Hearing	CDR
85	2019	The American Public Lands and Waters Climate Solution Act of 2019 (H. R. 5435)	Raúl Grijalva (Sponsor)	Proposed Bill (Introduced in House (IH))	CDR
86	2019	Congressional Record Volume 165, Number 199: Committee on Energy and Natural Resources	United States Senate	Congressional Record	CDR
87	2019	National Defense Authorization Act for Fiscal Year 2020 (to accompany S. 1790)	House of Representatives	Report	CDR
88	2019	Atmospheric Climate Intervention Research Act (H. R. 5519)	Jerry McNerney (sponsor)	Proposed Bill (Introduced in House (IH))	Umbrella term
89	2019	Climate Action Rebate Act of 2019 (H. R. 4051)	James Panetta, Thomas Suozzi (sponsors)	Proposed Legislation (Introduced in House (IH)	CDR
90	2020	Fossil Energy Research and Development Act of 2019 (H. R. 3607)	Marc Veasey and others (sponsors)	Proposed Legislation (Reported in House (RH))	CDR
91	2020	America's Transportation Infrastructure Act Of 2019 (to accompany S. 2302)	John Barrasso	Report	CDR
92	2020	Carbon Capture, Utilization, and Storage Innovation Act (CCUS Innovation Act) (H. R. 5865)	David McKinley (sponsor)	Proposed Bill (Introduced in House (IH))	CDR
93	2020	Clean Economy Act of 2020 (S. 3269)	Tom Carper and others (sponsor)	Proposed Bill (Introduced in Senate (IS))	CDR

	YEAR	TITLE	AUTHOR	DOC TYPE	CLIMATE ENGINEERING CONTEXT
94	2020	H.R. 5435, 'American Public Lands and Waters Climate Solution Act of 2019' and H.R. 5859, 'Trillion Trees Act"	Committee on Natural Resources, House of Representatives	Legislative Hearing	CDR
95	2020	Congressional Record Volume 166, Number 42: S. 2657	United States Senate	Congressional Record	CDR
96	2020	Congressional Record Volume 166, Number 119: S. 4049	United States Senate	Congressional Record	CDR
97	2020	Commerce, Justice, Science, And Related Agencies Appropriations Bill, 2021 (to accompany H.R. 7667)	José Serrano (Committee on Appropriations, House of Representatives)	Report	CDR
98	2020	Clean Energy Innovation and Deployment Act of 2020 (H. R. 7516)	Diana DeGette, Jared Huffman, Scott Peters (sponsors)	Proposed Bill (Introduced in House (IH))	CDR
99	2020	Defense, Commerce, Justice, Science, Energy and Water Development, Financial Services and General Government, Labor, Health and Human Services, Education, Transportation, Housing, and Urban Development Appropriations Act, 2021 (H. R. 7617)	Congress	Proposed Bill (Engrossed in House EH)	CDR
100	2020	Moving Forward Act (H.R. 2)	Congress	Proposed Bill (Engrossed in House EH)	CDR
101	2020	Energy and Water Development and Related Agencies Appropriations Bill 2021 (to accompany H.R. 7613)	Marcy Kaptur (Committee on Appropriations, House of Representatives)	Report	CDR
102	2020	America's Clean Future Fund Act (S. 4484)	Richard Durbin, United States Senate (Sponsor)	Proposed Bill (Introduced in Senate (IS))	CDR
103	2020	William M. (Mac) Thornberry National Defense Authorization Act for Fiscal Year 2021 (H. R. 6395)	Senate	Proposed Bill (Placed on Calendar Senate (PCS))	CDR

	YEAR	TITLE	AUTHOR	DOC TYPE	CLIMATE ENGINEERING CONTEXT
104	2020	Fossil Energy Research and Development Act of 2019 (to accompany H.R. 3607)	Eddie Johnson, Committee on Science, Space, and Technology, House of Representatives	Report	CDR
105	2020	Report 116–528 (to accompany H. Res. 1129)	Jim McGovern, Committee on Rules, House of Representatives	Report	CDR
106	2020	Clean Economy Jobs and Innovation Act (H. R. 4447)	Congress	Proposed Bill (Engrossed in House (EH))	CDR

REFERENCES

Abate, R. S., and A. B. Greenlee, 'Sowing Seeds Uncertain: Ocean Iron Fertilization, Climate Change, and the International Environmental Law Framework', *Pace Environmental Law Review*, 27 (2009): 555–98.
Abbott, A, *The System of Professions: An Essay on the Division of Expert Labor* (Chicago: University of Chicago Press, 2014).
Adger, W. N., J. Barnett, K. Brown, N. Marshall, and K. O'Brien, 'Cultural Dimensions of Climate Change Impacts and Adaptation', *Nature Climate Change*, 3.2 (2013): 112–17.
American Geophysical Union [AGU], *Geoengineering Solutions to Climate Change Require Enhanced Research, Consideration of Societal Impacts, and Policy Development* (Washington D.C., 13 September 2009).
Allan, B. B, 'Producing the Climate: States, Scientists, and the Constitution of Global Governance Objects', *International Organization*, 71.1 (2017): 131.
American Meteorological Society [AMS], *Geoengineering the Climate System. A Policy Statement of the American Meteorological Society* (Boston MA: American Meteorological Society, 6 January 2013).
Andrews, R., *Managing the Environment, Managing Ourselves: A History of American Environmental Policy* (New Haven: Yale University Press, 2006).
———, 'The EPA at 40: An Historical Perspective', *Duke Environmental Law and Policy Forum*, 21 (2010): 223–258.
Advanced Research Projects Agency – Energy [ARPA-E], 'ARPA-E Documentation', <https://arpa-e.energy.gov/?q=site-page/arpa-e-documentation> [accessed 3 March 2017].
Arrhenius, S., 'On the Influence of Carbonic Acid in the Air Upon the Temperature of the Ground', *The London, Edinburgh, and Dublin Philosophical Magazine and Journal of Science*, 41.251 (1896): 237–76.
———, *Worlds in the Making: The Evolution of the Universe* (New York, London: Harper & Brothers, 1908).
Asafu-Adjaye, J., L. Blomquist, S. Brand, B. Brook, R. DeFries, E. Ellis, and others, *An Ecomodernist Manifesto*, www.ecomodernism.org [accessed 15 May 2015].
Baatz, C., C. Heyward, and H. Stelzer, 'The Ethics of Engineering the Climate', *Environmental Values*, 25.1 (2016): 1–5.

Baker, Z., 'Climate State: Science-State Struggles and the Formation of Climate Science in the US from the 1930s to 1960s', *Social Studies of Science*, 47.6 (2017): 861–87.

Bala, G., and A. Gupta, 'Geoengineering and India', *Current Science*, 113.3 (2017): 376–77.

——, 'Solar Geoengineering Research in India', *Bulletin of the American Meteorological Society*, 100.1 (2019): 23–28.

Barnett, M., 'DOE Releases Climate Change Technology Program Strategic Plan', <https://energy.gov/articles/doe-releases-climate-change-technology-program-strategic-plan> [accessed 6 September 2006].

Barrett, S., 'The Incredible Economics of Geoengineering', *Environmental and Resource Economics*, 39.1 (2008): 45–54.

Beck, S., *Das Klimaexperiment und der IPCC: Schnittstellen Zwischen Wissenschaft Und Politik in Den Internationalen Beziehungen* (Marburg: Metropolis Verlag, 2009).

——, 'Moving Beyond the Linear Model of Expertise? IPCC and the Test of Adaptation', *Regional Environmental Change*, 11.2 (2011): 297–306.

——, Hybride Organisationen als Schnittstellen. Der Fall Des IPCC', in Besio, C. and G. Romano, eds, *Zum Gesellschaftlichen Umgang mit dem Klimawandel. Kooperationen und Kollisionen*, (Baden-Baden: Nomos, 2016).

Beck, S., and M. Mahony, 'The Politics of Anticipation: The IPCC and the Negative Emissions Technologies Experience', *Global Sustainability*, 1, online (2018).

Beck, U., *Risikogesellschaft: Auf Dem Weg in Eine Andere Moderne* (Frankfurt a.M.: Suhrkamp, 1986).

Bellamy, R., J. Chilvers, N. E. Vaughan, and T. M. Lenton, 'Appraising Geoengineering', *Tyndall Centre for Climate Change Research*, 2012.

Belter, C. W., and D. J. Seidel, 'A Bibliometric Analysis of Climate Engineering Research', *Wiley Interdisciplinary Reviews: Climate Change*, 4.5 (2013): 417–27.

Bickel, J. E., and L. Lane, 'An Analysis of Climate Engineering as a Response to Climate Change', *Copenhagen Consensus Center* (2009), <https://www.copenhagenconsensus.com/sites/default/files/ap_climate-engineering_bickel_lane_v.5.0.pdf> [accessed 18 August 2021].

Biermann, F., and I. Möller, 'Rich Man's Solution? Climate Engineering Discourses and the Marginalization of the Global South', *International Environmental Agreements: Politics, Law and Economics*, 2019.

Bimber, B. A., *The Politics of Expertise in Congress: The Rise and Fall of the Office of Technology Assessment* (Albany, NY: State University of New York Press, 1996).

Blackstock, J., and S. Low, eds, *Geoengineering Our Climate? Ethics, Politics, and Governance.* (New York: Routledge, 2019).

Bodansky, D., 'May We Engineer the Climate?', *Climatic Change*, 33.3 (1996): 309–21.

REFERENCES

Bonneuil, C., and J.-B. Fressoz, *The Shock of the Anthropocene: The Earth, History and Us* (New York: Verso Books, 2016).

Boswell, C., *The Political Uses of Expert Knowledge: Immigration Policy and Social Research* (Cambridge: Cambridge University Press, 2012).

Bourdieu, P., *Vom Gebrauch der Wissenschaft: für eine klinische Soziologie des wissenschaftlichen Feldes* (Konstanz: UVK, Universitäts Verlag, 1998).

———, *Science of Science and Reflexivity* (Chicago: University of Chicago Press, 2004).

Bracmort, K., and R. K. Lattanzio, *Geoengineering: Governance and Technology Policy (Update)* (Washington, D.C: Congressional Research Service, 2013).

Bracmort, K., R. K. Lattanzio, and E. C Barbour, *Geoengineering: Governance and Technology Policy* (Washington D.C.: Congressional Research Service, 2013).

Brand, S., *Whole Earth Discipline: Why Dense Cities, Nuclear Power, Transgenic Crops, Restored Wildlands, and Geoengineering Are Necessary* (London: Penguin Books, 2010).

Brechin, S. R., and D. A. Freeman, 'Public Support for Both the Environment and an Anti-Environmental President: Possible Explanations for the George W. Bush Anomaly', in *The Forum* (De Gruyter, 2004).

Bright, M., 'Getting Serious about Direct Air Capture', *Third Way*, 2020 <https://www.thirdway.org/memo/getting-serious-about-direct-air-capture> [accessed 29 October 2020].

Brooks, S., M. Leach, H. Lucas, and E. Millstone, *Silver Bullets, Grand Challenges and the New Philantropy*, STEPS Working Paper 24 (Brighton: STEPS Centre, 2009).

Brudnick, I. A., *The Congressional Research Service and the American Legislative Process* (Washington D.C.: Congressional Research Service, 2008).

Buck, H. J., 'Climate Engineering: Spectacle, Tragedy or Solution? A Content Analysis of News Media Framing', in *De-/Constructing the Greenhouse: Interpretative Approaches to Global Climate Governance. Routledge, London* (London: Routledge, 2013).

———, *After Geoengineering: Climate Tragedy, Repair, and Restoration* (New York City: Verso Books, 2019)

Budyko, M. I., *Climatic Changes* (American Geophysical Union, 1977).

Calvert, J., 'Systems Biology, Big Science and Grand Challenges', *BioSocieties*, 8.4 (2013): 466–79.

Cao, L., C. Gao, and L. Zhao, 'Geoengineering: Basic Science and Ongoing Research Efforts in China', *Advances in Climate Change Research*, 2015, <https://doi.org/10.1016/j.accre.2015.11.002> [accessed 14 May 2020].

Carlin, A., 'A Risky Gamble', *The Environmental Forum*, (2007): 42–47.

Chadwick, A., R. Arts, O. Eiken, P. Williamson, and G. Williams, 'Geophysical Monitoring of the CO2 Plume at Sleipner, North Sea', in S. Lombardi, L.

Altunina, S. Beaubien, eds, *Advances in the Geological Storage of Carbon Dioxide* (Dodrecht: Springer, 2006), pp. 303–14.

Chait, J., 'Obama Might Actually Be the Environmental President', *New York Magazine*, (2013) <https://nymag.com/news/features/obama-climate-change-2013-5/> [accessed 18 August 2021].

Crutzen, P. J., 'Albedo Enhancement by Stratospheric Sulfur Injections: A Contribution to Resolve a Policy Dilemma?', *Climatic Change*, 77.3 (2006): 211–20.

Czapla, E., 'ARPA-C, Same as ARPA-E', *American Action Forum* (2020), <https://www.americanactionforum.org/insight/arpa-c-same-as-arpa-e/> [accessed 7 July 2021].

Da-Allada, C. Y., E. Baloïtcha, E. A. Alamou, F. M. Awo, F. Bonou, Y. Pomalegni, and others, 'Changes in West African Summer Monsoon Precipitation under Stratospheric Aerosol Geoengineering', *Earth's Future* (2020), <https://doi.org/10.1029/2020EF001595> [accessed 18 August 2021].

Deutscher Bundestag, *Schriftliche Fragen Mit Den in Der Woche Vom 1. März 2010 Eingegangenen Antworten Der Bundesregierung*, Drucksache 19/27332 (2010), <https://dserver.bundestag.de/btd/19/273/1927332.pdf> [accessed 18 August 2021]

Dick, S., '*The Birth of NASA*' (NASA, 2008) <https://www.nasa.gov/exploration/whyweexplore/Why_We_29.html> [accessed 15 May 2018]

Die Deutsche Bundesregierung, *Antwort Der Bundesregierung Auf Die Kleine Anfrage Der Abgeordneten René Röspel, Dr. Ernst Dieter Rossmann, Oliver Kaczmarek, Weiterer Abgeordneter Und Der Fraktion Der SPD*, Drucksache 17/9943 (2012).

Dykema, J., D. Keith, J. Anderson, and D. Weisenstein, 'Stratospheric Controlled Perturbation Experiment: A Small-Scale Experiment to Improve Understanding of the Risks of Solar Geoengineering', *Philosophical Transactions of the Royal Society A: Mathematical, Physical and Engineering Sciences*, 372.2031 (2014).

Edney, K., and J. Symons, 'China and the Blunt Temptations of Geo-Engineering: The Role of Solar Radiation Management in China's Strategic Response to Climate Change', *The Pacific Review*, 27.3 (2014), 307–32 <https://doi.org/10.1080/09512748.2013.807865> [accessed 18 July 2020].

Edwards, P. N., 'Meteorology as Infrastructural Globalism', *Osiris*, 21.1 (2006): 229–50.

———, *A Vast Machine: Computer Models, Climate Data, and the Politics of Global Warming* (Cambridge, Massachusetts: MIT Press, 2010).

Ekholm, N., 'On the Variations of the Climate of the Geological and Historical Past and Their Causes', *Quarterly Journal of the Royal Meteorological Society*, 27.117 (1901): 1–62.

REFERENCES

Equinor, *'Sleipner Area'* (2019) <https://www.equinor.com/en/what-we-do/norwegian-continental-shelf-platforms/sleipner.html> [accessed 9 January 2019].

Eyal, G., 'For a Sociology of Expertise: The Social Origins of the Autism Epidemic', *American Journal of Sociology*, 118.4 (2013): 863–907.

——, *The Crisis of Expertise* (Cambridge: Polity Press, 2019).

Fialka, J., 'NOAA Gets Go-Ahead to Study Controversial Climate Plan B', *Scientific American*, 23 January 2020 <https://www.scientificamerican.com/article/noaa-gets-go-ahead-to-study-controversial-climate-plan-b/> [accessed 18 May 2020].

Fiekowsky, P., 'Restoring Climate Health Through Innovative Solutions', 2019 <https://www.commonwealthclub.org/events/2019-02-05/restoring-climate-health-through-innovative-solutions> [accessed 2 November 2019].

Fincham, M., 'The Day Before Yesterday: When Abrupt Climate Change Came to the Chesapeake Bay', *NOAA – Climate Governance* (2014), <https://www.climate.gov/print/181743> [accessed 2 November 2019].

Fleagle, R. G., 'NOAA's Role and the National Interest', *Science, Technology, & Human Values*, 11.2 (1986): 51–62.

Fleming, J., *Historical Perspectives on Climate Change* (Oxford: Oxford University Press, 1998).

——, 'The Pathological History of Weather and Climate Modification: Three Cycles of Promise and Hype', *Historical Stududies in the Physical and Biological Sciences*, 37.1 (2006): 3–25.

——, *Fixing the Sky: The Checkered History of Weather and Climate Control* (New York: Columbia University Press, 2010).

——, *Inventing Atmospheric Science: Bjerknes, Rossby, Wexler, and the Foundations of Modern Meteorology* (Cambridge, MA: MIT Press, 2016).

Fragniere, A., and S. Gardiner, 'Why Geoengineering Is Not "Plan B"', in C. Preston, ed., *Justice and Geoengineering* (Lanham: Rowman and Littlefield, 2016).

Gramelsberger, G., 'What Do Numerical (Climate) Models Really Represent?', *Studies in History and Philosophy of Science Part A*, 42.2 (2011): 296–302.

Gramelsberger, G., and J. Feichter, *Climate Change and Policy: The Calculability of Climate Change and the Challenge of Uncertainty* (Berlin, Heidelberg: Springer, 2011).

Grundmann, R., 'The Problem of Expertise in Knowledge Societies', *Minerva*, 55.1 (2017): 25–48.

Gupta, A., and I. Möller, 'De Facto Governance: How Authoritative Assessments Construct Climate Engineering as an Object of Governance', *Environmental Politics*, 28.3 (2019), 480–501.

Haas, P. M., 'Introduction: Epistemic Communities and International Policy Coordination', *International Organization*, 46.01 (1992): 1–35.

Hackmann, H., S. C. Moser, and A. L. St. Clair, 'The Social Heart of Global Environmental Change', *Nature Climate Change*, 4.8 (2014): 653–55.

Hale, B., 'The World That Would Have Been: Moral Hazard Arguments Against Geoengineering', in C. Preston, ed., *Engineering the Climate: The Ethics of Solar Radiation Management* (Lanham: Rowman and Littlefield, 2012).

Hamilton, C., *Earthmasters: The Dawn of the Age of Climate Engineering* (New Haven: Yale University Press, 2013).

Hannigan, J. A., *Environmental Sociology*, 2nd Edition (London, New York: Routledge, 2006).

Harper, K. C., *Weather by the Numbers: The Genesis of Modern Meteorology* (Cambridge, Mass.: MIT Press, 2008).

Hart, D. M., and D. G. Victor, 'Scientific Elites and the Making of US Policy for Climate Change Research, 1957-74', *Social Studies of Science*, 23.4 (1993): 643–80.

Heintz, B., 'Welterzeugung Durch Zahlen Modelle Politischer Differenzierung in Internationalen Statistiken, 1948–2010', *Soziale Systeme*, 18 (2012): 7–39.

Hezir, J., A. Stark, T. Bushman, and E. Smith, *Carbon Removal: Comparing Historical Federal Research Investments with the National Academies' Recommended Future Funding Levels* (Washington D.C.: Bipartisan Policy Center, April 2019), <https://bipartisanpolicy.org/wp-content/uploads/2019/06/Carbon-Removal-Comparing-Historical-Investments-with-the-National-Academies-Recommendations.pdf> [accessed 29 October 2020].

Hicks, D., 'Grand Challenges in US Science Policy Attempt Policy Innovation', *International Journal of Foresight and Innovation Policy*, 11.1/2/3 (2016): 22–42.

Hilgartner, S., *Science on Stage: Expert Advice as Public Drama* (Stanford, California: Stanford University Press, 2000).

Horton, J. B., 'The Emergency Framing of Solar Geoengineering: Time for a Different Approach', *The Anthropocene Review*, 2.2 (2015), 147–51.

Hulme, M., *What Sorts of Knowledge for What Sort of Politics? Science, Climate Change and the Challenge of Democracy* (Working Paper, Science, Society and Sustainability (3S) Research Group, University of East Anglia, Norwich, 2012).

——, *Can Science Fix Climate Change? A Case Against Climate Engineering* (Cambridge, Mass.: Polity Press, 2014).

Hurlbut, J. B., 'Reimagining Responsibility in Synthetic Biology', *Journal of Responsible Innovation*, 2.1 (2015): 113–16.

Huttunen, S., and M. Hildén, 'Framing the Controversial Geoengineering in Academic Literature', *Science Communication*, 36.1 (2014): 3–29.

REFERENCES

Huttunen, S., E. Skytén, and M. Hildén, 'Emerging Policy Perspectives on Geoengineering: An International Comparison', *The Anthropocene Review*, 2.1 (2015): 14–32.

International Maritime Organization [IMO], ed., 'United States Submission to the IMO Scientifc Group of the London Convention, "Planktos, Inc., Large-Scale Ocean Iron Addition Projects"', 2007.

Implications and Risks of Engineering Solar Radiation to Limit Climate Change [IMPLICC], 'Brief Summary of Scientific Results' (2019), <http://implicc.zmaw.de/index.php?id=551> [accessed 17 February 2021].

International Organization for Standardization [ISO], 'ISO/AWI TR 14082: Radiative Forcing Management – Guidance for the Quantification and Reporting of Radiative Forcing-Based Climate Footprints and Mitigation Efforts', <https://www.iso.org/standard/68505.html> [accessed 17 February 2021]

Intergovernmental Panel on Climate Change [IPCC], ed., *Climate Change. The IPCC Scientific Assessment, First Assessment Report* (Geneva, Switzerland: Intergovernmental Panel on Climate Change, 1990).

——, ed., *Climate Change 1995. Second Assessment Report of the Intergovernmental Panel on Climate Change* (Geneva, Switzerland: Intergovernmental Panel on Climate Change, 1995).

——, ed. 'Summary for Policymakers', in *Climate Change 2001: Mitigation. Contribution of Working Group III to the Third Assessment Report of the Intergovernmental Panel on Climate Change* (Geneva, Switzerland: Intergovernmental Panel on Climate Change, 2001)

——, ed., *IPCC Special Report on Carbon Dioxide Capture and Storage* (Geneva, Switzerland: Intergovernmental Panel on Climate Change, 2005)

——, ed., *Climate Change 2007: Mitigation of Climate Change: Contribution of Working Group III to the Fourth Assessment Report of the Intergovernmental Panel on Climate Change* (Geneva, Switzerland: Intergovernmental Panel on Climate Change, 2007)

——, ed., 'Summary for Policymakers', in *Climate Change 2013: The Physical Science Basis. Contribution of Working Group I to the Fifth Assessment Report of the Intergovernmental Panel on Climate Change* (Geneva, Switzerland: Intergovernmental Panel on Climate Change, 2013), pp. 1–30.

——, 'The Organization', 2017 <https://www.ipcc.ch/organization/organization.shtml> [accessed 19 October 2017]

——, ed., *Global Warming of 1.5°C An IPCC Special Report on the Impacts of Global Warming of 1.5°C above Pre-Industrial Levels and Related Global Greenhouse Gas Emission Pathways, in the Context of Strengthening the Global Response to the Threat of Climate Change, Sustainable Development, and Efforts to Eradicate*

Poverty (Geneva, Switzerland: Intergovernmental Panel on Climate Change, 2019)

Intergovernmental Panel on Climate Change [IPCC], United Nations Environment Program, and James G. Titus, *Changing Climate and the Coast. Volume 2: Western Africa, the Americas, the Mediterranean Basin, and the Rest of Europe*, May 1990.

Izrael, Y. A., V.M. Zakharov, N. N Petrov, A.G. Ryaboshapko, V.N. Ivanov, A.V. Savchenko, and others, 'Field Studies of a Geo-Engineering Method of Maintaining a Modern Climate with Aerosol Particles', *Russian Meteorology and Hydrology*, 34.10 (2009b): 635–38.

———, 'Field Experiment on Studying Solar Radiation Passing through Aerosol Layers', *Russian Meteorology and Hydrology*, 34.5 (2009a): 265–73.

Jacques, P. J., R. E. Dunlap, and M. Freeman, 'The Organisation of Denial: Conservative Think Tanks and Environmental Scepticism', *Environmental Politics*, 17.3 (2008): 349–85.

Jobst, C., 'Sozialwissenschaftliche Analyse von Klimaforschung, -diskurs und -politik am Beispiel des IPCC', in M. Voss, ed., *Der Klimawandel* (Wiesbaden: VS Verlag für Sozialwissenschaften, 2010), pp. 101–15.

Kaiser, F. M., *GAO: Government Accountability Office and General Accounting Office* (Washington D.C.: Congressional Research Service, 2007).

Kaldewey, D., 'The Grand Challenges Discourse: Transforming Identity Work in Science and Science Policy', *Minerva*, (2018): 161–82.

Karami, K., S. Tilmes, H. Muri, and S. V. Mousavi, 'Storm Track Changes in the Middle East and North Africa under Stratospheric Aerosol Geoengineering', *Geophysical Research Letters*, 2020.

Katz, E., 'Geoengineering, Restoration, and the Construction of Nature: Oobleck and the Meaning of Solar Radiation Management', *Environmental Ethics*, 37.4 (2015): 485–98.

Keeling, C. D., 'Rewards and Penalties of Monitoring the Earth', *Annual Review of Energy and the Environment*, 23.1 (1998): 25–82.

Keith, D., 'Geoengineering the Climate: History and Prospect', *Annual Review of Energy and the Environment*, 25.1 (2000): 245–84.

———, *A Case for Climate Engineering* (Cambridge, Mass.: MIT Press, 2013).

Keller, A. C., *Science in Environmental Policy: The Politics of Objective Advice* (Cambridge, Mass.: MIT Press, 2009).

Kellogg, W. W., and Stephen H. Schneider, 'Climate Stabilization: For Better or for Worse', *Science*, 186.4170 (1974): 1163–72.

Kincaid, G., and J. T. Roberts, 'No Talk, Some Walk: Obama Administration First-Term Rhetoric on Climate Change and US International Climate Budget Commitments', *Global Environmental Politics*, 13.4 (2013): 41–60.

REFERENCES

King, D., 'Climate Repair', *The Cambridge Climate Change Lecture Series*, 2019 <https://www.eventbrite.co.uk/e/climate-repair-tickets-55763880267> [accessed 2 November 2019].

Kintisch, E., *Hack the Planet: Science's Best Hope-or Worst Nightmare-for Averting Climate Catastrophe* (Hoboken: John Wiley & Sons, 2010).

Kravitz, B., K. Caldeira, O. Boucher, A. Robock, P. J. Rasch, K. Alterskjaer, and others, 'Climate Model Response from the Geoengineering Model Intercomparison Project (GeoMIP)', *Journal of Geophysical Research: Atmospheres*, 118.15 (2013): 8320–32.

Kravitz, B., D. G. MacMartin, D. Visioni, O. Boucher, J. N.S. Cole, J. Haywood, and others, 'Comparing Different Generations of Idealized Solar Geoengineering Simulations in the Geoengineering Model Intercomparison Project (GeoMIP)', *Atmospheric Chemistry and Physics Discussions* (2020): 1–31.

Kravitz, B., A. Robock, O. Boucher, H. Schmidt, K. E. Taylor, G. Stenchikov, and others, 'The Geoengineering Model Intercomparison Project (GeoMIP)', *Atmospheric Science Letters*, 12.2 (2011): 162–67.

Kravitz, B., A. Robock, S. Tilmes, O. Boucher, J. M. English, P. J. Irvine, and others, 'The Geoengineering Model Intercomparison Project Phase 6 (GeoMIP6): Simulation Design and Preliminary Results', *Geoscientific Model Development*, 8.10 (2015): 3379–92.

Kreuter, J., 'Technofix, Plan B or Ultima Ratio? A Review of the Social Science Literature on Climate Engineering Technologies', *Oxford University, Institute of Science, Innovation and Society*, Occasional Paper Series, 2 (2015).

Lahsen, M., 'Experiences of Modernity in the Greenhouse: A Cultural Analysis of a Physicist "Trio" Supporting the Backlash Against Global Warming', *Global Environmental Change*, 18.1 (2008): 204–19.

Latham, J., K. Bower, T. Choularton, H. Coe, P. Connolly, G. Cooper, and others, 'Marine Cloud Brightening', *Philosophical Transactions of the Royal Society A: Mathematical, Physical and Engineering Sciences*, 370.1974 (2012): 4217–62.

Latour, B., *Science in Action: How to Follow Scientists and Engineers through Society* (Cambridge, Mass.: Harvard University Press, 1987).

——, 'On Recalling ANT', *The Sociological Review*, 47.1_suppl (1999): 15–25.

Latour, B., and S. Woolgar, *Laboratory Life: The Construction of Scientific Facts* (New Jersey: Princeton University Press, 1979).

Lattanzio, R. K., and E. C. Barbour, 'Memorandum: International Governance of Geoengineering' (Washington D.C.: Congressional Research Service, 2010).

Lewis, J., 'The Birth of EPA', *EPA Journal*, 11 (1985).

Lin, A. C., 'Does Geoengineering Present a Moral Hazard', *Ecology LQ*, 40 (2013): 673–712.

van der Linden, S. L., A. A. Leiserowitz, G. D. Feinberg, and E. W. Maibach, 'How to Communicate the Scientific Consensus on Climate Change: Plain Facts, Pie Charts or Metaphors?', *Climatic Change*, 126.1–2 (2014): 255–62.

Linnér, B. O., and V. Wibeck, 'Dual High-Stake Emerging Technologies: A Review of the Climate Engineering Research Literature', *Wiley Interdisciplinary Reviews: Climate Change* (2015): 255–68.

Locher, F., and J.-B. Fressoz, 'Modernity's Frail Climate: A Climate History of Environmental Reflexivity', *Critical Inquiry*, 38.3 (2012): 579–98.

Long, J., S. Rademaker, J. Anderson, R. E. Benedick, K. Caldeira, J. Chaisson, and others, *Geoengineering: A National Strategic Plan for Research on the Potential Effectiveness, Feasibility, and Consequences of Climate Remediation Technologies* (The Bipartisan Policy Center, 2011).

Luhmann, N., *Die Wissenschaft Der Gesellschaft* (Frankfurt am Main: Suhrkamp, 1990).

———, *Die Politik Der Gesellschaft* (Frankfurt am Main: Suhrkamp, 2002)

———, *Theory of Society* (Stanford, California: Stanford University Press, 2013).

———, *Politische Soziologie*, ed. by A. Kieserling, 1. Aufl (Berlin: Suhrkamp, 2015).

Lukacs, M., S. Goldenberg, and A. Vaughan, 'Russia Urges UN Climate Report to Include Geoengineering', *The Guardian* (2013) <https://www.theguardian.com/environment/2013/sep/19/russia-un-climate-report-geoengineering> [accessed 17 February 2021].

Luokkanen, M., S. Huttunen, and M. Hildén, 'Geoengineering, News Media and Metaphors: Framing the Controversial', *Public Understanding of Science*, 23.8 (2014): 966–81.

Maasen, S., and P. Weingart, *Democratization of Expertise?: Exploring Novel Forms of Scientific Advice in Political Decision-Making* (Dodrecht: Springer, 2006).

MacDonald, G. J.F., and others, *The Long Term Impact of Atmospheric Carbon Dioxide on Climate, JASON Technical Report* (Arlington, Virginia: JASON, April 1979).

Malm, A., *Fossil Capital: The Rise of Steam Power and the Roots of Global Warming* (New York: Verso Books, 2016).

Marchetti, C., 'On Geoengineering and the CO2 Problem', *Climatic Change*, 1.1 (1977): 59–68.

Markusson, N., M. D. Gjefsen, J. C. Stephens, and D. Tyfield, 'The Political Economy of Technical Fixes: A Case from the Climate Domain', *Energy Research & Social Science*, 23 (2017): 1–10.

Markusson, N., F. Ginn, N. S. Ghaleigh, and V. Scott, '"In Case of Emergency Press Here": Framing Geoengineering as a Response to Dangerous Climate Change', *Wiley Interdisciplinary Reviews: Climate Change*, 5.2 (2014): 281–90.

Marland, G., 'Could We/Should We Engineer the Earth's Climate?', *Climatic Change*, 33.3 (1996): 275–78.

Massachusetts Institute of Technology [MIT], 'Sleipner Fact Sheet: Carbon Dioxide Capture and Storage Project', (2019), <https://sequestration.mit.edu/tools/projects/sleipner.html> [accessed 9 January 2019].

McCright, A. M., and R. E. Dunlap, 'Defeating Kyoto: The Conservative Movement's Impact on US Climate Change Policy', *Social Problems*, 50.3 (2003): 348–73.

——, 'The Politicization of Climate Change and Polarization in the American Public's Views of Global Warming, 2001–2010', *The Sociological Quarterly*, 52.2 (2011): 155–94.

McLaren, D., 'Power and Responsibility in a Broken World', *Forum for Climate Engineering Assessment*, (2018a), <http://ceassessment.org/climate-repair/> [accessed 8 August 2018].

——, 'In a Broken World: Towards an Ethics of Repair in the Anthropocene', *The Anthropocene Review*, 5.2 (2018b): 136–54.

Meadows, D. H. and Club of Rome, eds., *The Limits to Growth. A Report for the Club of Rome's Project on the Predicament of Mankind* (New York: Universe Books, 1972).

Miller, C., 'Hybrid Management: Boundary Organizations, Science Policy, and Environmental Governance in the Climate Regime', *Science, Technology & Human Values*, 26.4 (2001): 478–500.

——, 'Climate Science and the Making of a Global Political Order', in S. Jasanoff, ed., *States of Knowledge: The Co-Production of Science and Social Order* (London: Routlege, 2004), pp. 64–66.

Miller, C., and P. Edwards, eds., *Changing the Atmosphere: Expert Knowledge and Environmental Governance* (Cambridge, Mass.: MIT Press, 2001).

Mitchell, T., 'Fixing the Economy', *Cultural Studies*, 12.1 (1998): 82–101.

——, *Rule of Experts: Egypt, Techno-Politics, Modernity* (Berkeley: University of California Press, 2002).

Möller, I., *Potential Obstruction of Climate Change Mitigation through ISO Standard on Radiative Forcing Management* (Climate Social Science Network, 2021).

Moniz, E., J. Hezir, C. McCormick, T. Bushman, and S. Savitz, *Clearing the Air: A Federal RD&D Initiative and Management Plan for Carbon Dioxide Removal Technologies* (Washington D.C.: Energy Futures Initiative, September 2019).

Moore, J. C., Y. Chen, X. Cui, W. Yuan, W. Dong, Y. Gao, and others, 'Will China Be the First to Initiate Climate Engineering?', *Earth's Future*, 4.12 (2016): 588–95.

Morrow, D., 'When Technologies Makes Good People Do Bad Things: Another Argument Against the Value-Neutrality of Technologies', *Science and Engineering Ethics*, 20.2 (2014): 329–43.

——, 'International Governance of Climate Engineering: A Survey of Reports on Climate Engineering, 2009–2015', *SSRN Electronic Journal*, 2017 <https://doi.org/10.2139/ssrn.2982392> [accessed 18 September 2020].

Morseletto, P., F. Biermann, and P. Pattberg, 'Governing by Targets: Reductio Ad Unum and Evolution of the Two-Degree Climate Target', *International Environmental Agreements: Politics, Law and Economics*, 5.17 (2016): 655–76.

Morton, O., *The Planet Remade: How Geoengineering Could Change the World* (New Jersey: Princeton University Press, 2016).

Moser, S. C., 'Communicating Climate Change: History, Challenges, Process and Future Directions', *Wiley Interdisciplinary Reviews: Climate Change*, 1.1 (2010): 31–53.

Mukerji, C., *A Fragile Power: Scientists and the State* (Ney Jersey: Princeton University Press, 2014).

National Aeronautics and Space Agency [NASA], 'About NASA's History Division', 2017a <https://history.nasa.gov/program.html> [accessed 21 February 2017].

———, 'NIAC – NASA Institute for Advanced Concepts', 2017b <http://www.niac.usra.edu> [accessed 21 February 2017].

Necheless, E., L. Burns, A. Chang, and D. Keith, 'Funding for Solar Geoengineering from 2008 to 2018', *Harvard's Solar Geoengineering Research Program*, 2018 <https://geoengineering.environment.harvard.edu/blog/funding-solar-geoengineering> [accessed 29 October 2020].

Nerlich, B., and R. Jaspal, 'Metaphors We Die By? Geoengineering, Metaphors, and the Argument from Catastrophe', *Metaphor and Symbol*, 27.2 (2012): 131–47.

Nierenberg, N., W. R. Tschinkel, and V. J. Tschinkel, 'Early Climate Change Consensus at the National Academy: The Origins and Making of "Changing Climate"', *Historical Studies in the Natural Sciences*, 40.3 (2010): 318–49.

National Oceanic and Atmospheric Administration [NOAA], *Geophysical Monitoring for Climatic Change No. 1, Summary Report 1972*, Summary Reports (Boulder, Colorado, 1974),

———, *Ocean Fertilization: The Potential of Ocean Fertilization for Climate Change Mitigation. A Report to Congress*, 2010.

———, 'Our History', 2017 <http://www.noaa.gov/our-history> [accessed 21 February 2017].

Nordhaus, W., 'Greenhouse Economics: Count Before You Leap', *The Economist* (London, England, 7 July 1990).

———, 'An Optimal Transition Path for Controlling Greenhouse Gases', *Science*, 258.5086 (1992): 1315–19.

Oceanos, 'Ocean Seeding: Scientific Research', 2018 <http://oceaneos.org/ocean-seeding/ocean-fertilization-scientific-research/> [accessed 20 May 2018].

Office of Senator Whitehouse, 'Senators Call Out Web of Denial Blocking Action On Climate Change', *Sheldon Whitehouse*, 2016 <https://www.whitehouse.senate.

gov/news/release/senators-call-out-web-of-denial-blocking-action-on-climate-change> [accessed 20 May 2018].

Oldham, P., B. Szerszynski, J. Stilgoe, C. Brown, B. Eacott, and A. Yuille, 'Mapping the Landscape of Climate Engineering', *Philosophical Transactions of the Royal Society of London A: Mathematical, Physical and Engineering Sciences*, 372.2031 (2014): 1–20.

Owen, R., 'Solar Radiation Management and the Governance of Hubris', in R. E. Hester and R. M. Harrison, eds. *Geoengineering of the Climate System*, Issues in Environmental Science and Technology, 38 (2014): 212–214.

Paglia, E., 'The Socio-Scientific Construction of Global Climate Crisis', *Geopolitics*, 23.1 (2018): 96–123.

Petersen, A., 'The Emergence of the Geoengineering Debate Within the IPCC', *Working Paper*, Geoengineering Our Climate Working Paper and Opinion Article Series, 2014.

Peterson, A., 'Final FY20 Appropriations: DOE Applied Energy R&D', *American Institute of Physics*, 2020 <https://www.aip.org/fyi/2020/final-fy20-appropriations-doe-applied-energy-rd> [accessed 29 October 2020].

Pickersgill, M., 'Connecting Neuroscience and Law: Anticipatory Discourse and the Role of Sociotechnical Imaginaries', *New Genetics and Society*, 30.1 (2011): 27–40.

Pielke, R. A., 'Policy History of the US Global Change Research Program: Part I. Administrative Development', *Global Environmental Change*, 10.1 (2000a): 9–25.

———, 'Policy History of the US Global Change Research Program: Part II. Legislative Process', *Global Environmental Change*, 10.2 (2000b): 133–44.

———, *The Climate Fix: What Scientists and Politicians Won't Tell You about Global Warming* (New York: Basic Books, 2010).

Pierce, J. R., D. K. Weisenstein, P. Heckendorn, T. Peter, and D. W. Keith, 'Efficient Formation of Stratospheric Aerosol for Climate Engineering by Emission of Condensible Vapor from Aircraft', *Geophysical Research Letters*, 37.18 (2010).

Pierrehumbert, R., 'Climate Hacking Is Barking Mad', *Slate*, 2015.

Pinch, T. J., and W. E. Bijker, 'The Social Construction of Facts and Artefacts: Or How the Sociology of Science and the Sociology of Technology Might Benefit Each Other', *Social Studies of Science*, 14.3 (1984): 399–441.

Pinto, I., C. Jack, C. Lennard, S. Tilmes, and R. C. Odoulami, 'Africa's Climate Response to Solar Radiation Management With Stratospheric Aerosol', *Geophysical Research Letters*, 47.2 (2020).

Poloni, V., 'Das Intergovernmental Panel on Climate Change (IPCC) Als Boundary Organization', in *Organisationen Der Forschung* (Springer, 2009), pp. 250–71

Pontecorvo, E., 'The Climate Policy Milestone That Was Buried in the 2020 Budget', *GRIST*, 2020 <https://grist.org/climate/the-climate-policy-milestone-that-was-buried-in-the-2020-budget/> [accessed 29 October 2020].

Porter, T. M., 'Speaking Precision to Power: The Modern Political Role of Social Science', *Social Research: An International Quarterly*, 73.4 (2006): 1273–94.

Public Law 114–113, 'Consolidated Appropriations Act of 2016', 129 Stat. 2242; Date 18 December 2015; enacted H.R.2029 (Washington D.C.: US Congress 2015).

Public Law 110–69, 'America COMPETES Act', 121 Stat. 572; Date 9 August 2007; enacted H.R.2272 (Washington D.C.: US Congress 2007).

Rahman, A. Atiq, P. Artaxo, A. Asrat, and A. Parker, 'Developing Countries Must Lead on Solar Geoengineering Research', *Nature*, 556.7699 (2018): 22–24.

RAND Corporation, 'Weather Modification Progress and the Need for Interactive Research', *Bulletin of the American Meteorological Society*, Weather Modification Research Project; Santa Monica, 50 (1969): 216–46.

Readfearn, G., 'Artificial Fog and Breeding Coral: Study Picks Best Great Barrier Reef Rescue Ideas', *The Guardian*, 15 April 2020a <https://www.theguardian.com/environment/2020/apr/16/brightening-clouds-and-coral-larvae-study-picks-best-great-barrier-reef-rescue-ideas> [accessed 17 February 2021].

———, 'Scientists Trial Cloud Brightening Equipment to Shade and Cool Great Barrier Reef', *The Guardian*, 16 April 2020b <https://www.theguardian.com/environment/2020/apr/17/scientists-trial-cloud-brightening-equipment-to-shade-and-cool-great-barrier-reef> [accessed 17 February 2021].

Robock, A., '20 Reasons Why Geoengineering May Be a Bad Idea', *Bulletin of the Atomic Scientists*, 64.2 (2008), 14–18

———, 'Benefits and Risks of Stratospheric Solar Radiation Management for Climate Intervention (Geoengineering)', ed. by National Academy of Engineering NAE, *The Bridge*, Vol. 50.1 (2020): 59–68.

Rodney and Otamatea Times, Waitemata and Kaipara Gazette, 'Coal Consumption Affecting Climate' (14 August 1912).

Roman, M., and M. Carson, *Sea Change: US Climate Policy Prospects under the Obama Administration* (Stockholm: The Commission on Sustainable Development, 2009).

Royal Society, *Geoengineering the Climate: Science, Governance and Uncertainty* (London: The Royal Society, 2009).

———, 'About the Royal Society', 2017a <https://royalsociety.org/about-us/> [accessed 19 October 2017].

———, 'Mission and Priorities', 2017b <https://royalsociety.org/about-us/mission-priorities/> [accessed 19 October 2017].

REFERENCES

Russell, L. M., 'Offsetting Climate Change by Engineering Air Pollution to Brighten Clouds', *Bridge*, 42.4 (2012): 10–15.

Russell, L. M., A. Sorooshian, J. H. Seinfeld, B. A. Albrecht, A. Nenes, L. Ahlm, and others, 'Eastern Pacific Emitted Aerosol Cloud Experiment', *Bulletin of the American Meteorological Society*, 94.5 (2013): 709–29.

Sagarin, R., M. Dawson, D. Karl, A. Michael, B. Murray, M. Orbach, and others, 'Iron Fertilization in the Ocean for Climate Mitigation: Legal, Economic, and Environmental Challenges', *Nichols School of the Environment, Duke University*, Working Paper (2007), <https://nicholasinstitute.duke.edu/oceans/marinees/iron-fertilization-in-the-ocean-for-climate-mitigation-legal-economic-and-environmental-challenges> [accessed 18 August 2021].

Sarewitz, D., 'How Science Makes Environmental Controversies Worse', *Environmental Science & Policy*, 7.5 (2004): 385–403.

Sarewitz, D., and R. Pielke Jr., 'The Neglected Heart of Science Policy: Reconciling Supply of and Demand for Science', *Environmental Science & Policy*, 10.1 (2007): 5–16.

Schelling, T. C., 'The Economic Diplomacy of Geoengineering', *Climatic Change*, 33.3 (1996): 303–7.

Schneider, S. H., 'Geoengineering: Could? Or Should? We Do It?', *Climatic Change*, 33.3 (1996): 291–302.

Schubert, J., 'Measuring, Modelling, Controlling the Climate? Numerical Expertise in US Climate Engineering Politics', in M. Prutsch, ed., *Science, Numbers and Politics* (Basingstoke: Palgrave Macmillan, 2019): pp. 181–202.

———, 'Die Politische Wirkmacht Wissenschaftlicher Expertise. Natürliche Analogien und Theoretische Experimente in US Amerikanischer Geoengineering Politik', in U. Büttner and D. Müller, eds, *Dritte Natur: Climate Engineering*, 3, (Berlin: Matthes und Seitz, 2021).

Schwarber, A., 'Final FY20 Appropriations: National Oceanic and Atmospheric Administration', *American Institute of Physics*, 2020a <https://www.aip.org/fyi/2020/final-fy20-appropriations-national-oceanic-and-atmospheric-administration> [accessed 29 October 2020].

———, 'FY21 Budget Outlook: National Oceanic and Atmospheric Administration', *American Institute of Physics*, 2020b <https://www.aip.org/fyi/2020/fy21-budget-outlook-national-oceanic-and-atmospheric-administration> [accessed 29 October 2020].

Scott, J. C., *Seeing like a State: How Certain Schemes to Improve the Human Condition Have Failed* (New Haven: Yale University Press, 1998).

Shabecoff, P., 'Global Warming Has Begun, Expert Tells Senate', *The New York Times* (New York, 24 June 1988).

Sillmann, J., T. M. Lenton, A. Levermann, K. Ott, M. Hulme, F. Benduhn, and others, 'Climate Emergencies Do Not Justify Engineering the Climate', *Nature Climate Change*, 5.4 (2015): 290–92.

SilverLining, 'SilverLining Announces $3 Million Safe Climate Research Initiative Supporting Research on Rapid Climate Interventions', 2020 <https://static1.squarespace.com/static/5bbac81c7788975063632c65/t/5f9973d740e38c75e7c14988/1603892184077/Safe+Climate+Research+Initiative+Press+Release+Formatted.pdf> [accessed 3 November 2020].

Sismondo, S., 'Models, Simulations, and Their Objects', *Science in Context*, 12.02 (1999): 247–60.

Smallman, M., '"Nothing to Do with the Science": How an Elite Sociotechnical Imaginary Cements Policy Resistance to Public Perspectives on Science and Technology through the Machinery of Government', *Social Studies of Science*, 50.4 (2020): 589–608.

Specter, M., 'The Climate Fixers', *New Yorker*, 2012 <http://www.newyorker.com/magazine/2012/05/14/the-climate-fixers> [accessed 13 August 2014].

SPICE, Stratospheric Particle Injection for Climate Engineering, 'SPICE. Aims and Background', 2018 <http://www.spice.ac.uk/about-us/aims-and-background/> [accessed 13 August 2018].

Stichweh, R., 'Gelehrter Rat und wissenschaftliche Politikberatung', in G. Putlitz, ed., *Politikberatung in Deutschland* (Wiesbaden: VS Verlag für Sozialwissenschaften, 2006), pp. 101–12.

———, 'Analysing Linkages Between Science and Politcs. Transformations of Functional Differentiation in Contemporary Society', in Stiftung Mercator, ed., *Interfaces of Science and Policy and the Role of Foundations* (Stiftung Mercator, 2015): pp. 38–47.

Stilgoe, J., *Experiment Earth: Responsible Innovation in Geoengineering* (London, New York: Routledge, Taylor & Francis Group, 2015).

Study of Critical Environmental Problems [SCEP], *Man's Impact on the Global Environment: Assessment and Recommendations for Action* (Cambridge, Massachusetts: MIT Press, 1970).

Taylor, P. J., and F. H. Buttel, 'How Do We Know We Have Global Environmental Problems? Science and the Globalization of Environmental Discourse', *Geoforum*, 23.3 (1992): 405–16.

Temple, J., 'China Builds One of the World's Largest Geoengineering Research Programs', *MIT Technology Review*, 2017 <https://www.technologyreview.com/s/608401/china-builds-one-of-the-worlds-largest-geoengineering-research-programs/> [accessed 29 October 2020].

———, 'The US Government Has Approved Funds for Geoengineering

Research', *MIT Technology Review* (2019) <https://www.technologyreview.com/2019/12/20/131449/the-us-government-will-begin-to-fund-geoengineering-research/> [accessed 29 October 2020].

The Biden Harris Campaign, 'The Biden Plan to Build a Modern, Sustainable Infrastructure and an Equitable Clean Energy Future', *Biden Harris Campaign* (2020) <https://joebiden.com/clean-energy/> [accessed 20 February 2021].

Tilmes, S., J. Fasullo, J.-F. Lamarque, D. R. Marsh, M. Mills, K. Alterskjaer, and others, 'The Hydrological Impact of Geoengineering in the Geoengineering Model Intercomparison Project (GeoMIP)', *Journal of Geophysical Research: Atmospheres*, 118.19 (2013): 11–36.

Tollefson, J., 'First Sun-Dimming Experiment Will Test a Way to Cool Earth', *Nature News Feature*, 2018.

Turner, J. M., and A. Isenberg, *The Republican Reversal: Conservatives and the Environment from Nixon to Trump* (Cambridge, Mass.: Harvard University Press, 2018).

U.K. House of Commons, Science and Technology Committee, *The Regulation of Geoengineering. Fifth Report of Session 2009–10* (London: The Stationery Office Limited, 03 2010).

Umwelt Bundesamt, *Geoengineering. Wirksamer Klimaschutz Oder Größenwahn?* (Dessau-Roßlau: Umweltbundesamt, 2011).

US Climate Change Technology Program [CCTP], *Strategic Plan* (Washington, D.C., 2006).

US Department of Energy [DOE], '69 FR 21514: Notice of Intent to Prepare a Programmatic Environmental Impact Statement for Implementation of the Carbon Sequestration Program' (Washington, D.C.: US Government Printing Office, 2004).

———, 'Carbon Storage Research', (2017), <https://www.energy.gov/fe/science-innovation/carbon-capture-and-storage-research> [accessed 20 February 2017].

US Department of Energy [DOE], Office of Fossil Energy, 'DOE Study Monitors Carbon Dioxide Storage in Norway's Offshore Sleipner Gas Field', *US Department of Energy*, (2009), <https://www.energy.gov/fe/articles/doe-study-monitors-carbon-dioxide-storage-norways-offshore> [accessed 20 February 2017].

US Department of Justice, Antitrust Division, '68 FR 16552: Notice Pursuant to the National Cooperative Research and Production Act of 1993 – Global Climate and Energy Project' (US Government Printing Office, 2003).

US Global Change Research Program [USGCRP], *Climate Science Special Report: Fourth National Climate Assessment: Volume I* (Washington D.C.: US Global Change Research Program, 2017).

US Global Climate and Energy Project [GCEP], 'Stanford University Global Climate and Energy Project. About Us', (2017a), <http://gcep.stanford.edu/about/index.html> [accessed 20 February 2017].

——, 'Stanford University Global Climate and Energy Project: FAQs', (2017b), <http://gcep.stanford.edu/about/faqs.html> [accessed 20 February 2017].

US Government Accountability Office [GAO], *Climate Change: Preliminary Observations on Geoengineering Science, Federal Efforts, and Governance Issues* (Washington, D.C.: US Government Accountability Office, March 2010a).

——, *Climate Change: A Coordinated Strategy Could Focus Federal Geoengineering Research and Inform Governance Efforts: Report to the Chairman, Committee on Science and Technology, House of Representatives* (Washington, D.C.: US Government Accountability Office, 2010b).

——, *Climate Engineering: Technical Status, Future Directions and Potential Responses* (Washington D.C.: US Government Accountability Office, 2011)

US Government Publishing Office [GPO], 'FDsys: Collections', (2018) <https://www.gpo.gov/help/index.html#what_s_available.htm> [accessed 24 May 2018].

US Environmental Protection Agency [EPA], '75 FR 77229: Final Rule on Federal Requirements Under the Underground Injection Control (UIC) Program for Carbon Dioxide Geologic Sequestration (GS) Wells' (US Government Printing Office, 2010).

——, '80 FR 64661: Final Rule on Carbon Pollution Emission Guidelines for Existing Stationary Sources: Electric Utility Generating Units' (Washington, D.C.: US Government Printing Office, 2015a).

——, '80FR 64966: Proposed Rule on Federal Plan Requirements for Greenhouse Gas Emissions from Electric Utility Generating Units Constructed on or Before January 8, 2014; Model Trading Rules; Amendments to Framework Regulations' (Washington, D.C.: US Government Printing Office, 2015b).

US House of Representatives, 111th Congress, '*H.R.2454 – American Clean Energy and Security Act of 2009*', Version 7 July 2009, (Washington, D.C.: US Government Printing Office, 2009).

——, '*Commerce, Justice, Science, and Related Agencies Appropriations Bill, 2010: Report Together with Additional Views to Accompany H.R. 2847*' (Washington, D.C.: US Government Printing Office, 12 June 2009).

US House of Representatives, 112th Congress, '*Commerce, Justice, Science, and Related Agencies Appropriations Bill, 2012: Report Together with Minority Views to Accompany H.R. 2596*' (Washington, D.C.: US Government Printing Office, 2012).

US House of Representatives, 115th Congress, '*H.R.4586 – Geoengineering Research*

Evaluation Act of 2017', Version 7 December 2017 (Washington, D.C.: US Government Printing Office: 2017).

US House of Representatives, 116th Congress, *'H.R.3607 – Fossil Energy Research and Development Act of 2019'*, Version 2 July 2019 (Washington, D.C.: US Government Printing Office: 2019).

———, *'H.R.5519 – Atmospheric Climate Intervention Act'*, Version 19 December 2019 (Washington, D.C.: US Government Printing Office: 2019).

US House of Representatives, Committee on Appropriations, *Commerce, Justice, Science, and Related Acencies Appropriations Act, 2020*, Report (Washington, D.C.: US Government Printing Office: 2019).

US House of Representatives, Committee on Energy and Commerce, *'Climate Leadership and Environmental Action for Our Nation's (CLEAN) Future Act*, Discussion Draft (Washington, D.C.: US Government Printing Office, 2020).

US House of Representatives, Committee on Energy Independence and Global Warming, *Building US Resilience to Global Warming Impacts*, Hearing (Washington, D.C.: US Government Printing Office, 2009).

US House of Representatives, Committee on Government Reform, *Climate Change Technology Research: Do We Need a 'Manhattan Project' for the Environment?*, Hearing (Washington, D.C.: US Government Printing Office, 2006).

US House of Representatives, Committee on Science and Technology, *Geoengineering: Parts I, II, and III*, Hearing (Washington, DC: US Government Printing Office, 2009).

———, *Fiscal Year 2011 Research and Development Budget Proposals at the Environmental Protection Agency (EPA) and the National Oceanic and Atmospheric Administration (NOAA)*, Hearing (Washington, D.C.: US Government Printing Office, 2010a).

———, *Engineering the Climate: Research Needs and Strategies for International Coordination*, Committee Print (Washington, D.C.: US Government Printing Office, October 2010b).

US House of Representatives, Committee on Science, Space, and Technology, *Geoengineering: Innovation, Research, and Technology*, Hearing (Washington D.C.: US Government Printing Office, 2017).

US House of Representatives, Select Committee on Energy Independence and Global Warming, *Voluntary Carbon Offsets: Getting What You Pay For*, Hearing (Washington, D.C.: US Government Printing Office, 2007).

———, *Not Going Away: America's Energy Security, Jobs and Climate Challenges*, Hearing (Washington, D.C.: US Government Printing Office, 2010).

US House of Representatives, Select Committee on the Climate Crisis, *Solving the Climate Crisis: The Congressional Action Plan for a Clean Energy Economy and a*

Healthy, Resilient, and Just America, Report (Washington D.C.: US Government Printing Office, 2020).

US House of Representatives, Subcommittee on Africa and Global Health of the Committee on Foreign Affairs, *Combating Climate Change in Africa*, Hearing (Washington, D.C.: US Government Printing Office, 2010).

US House of Representatives, Subcommittee on Commerce, Justice, Science, and Related Agencies of the Committee on Appropriations, *Commerce, Justice, Science, and Related Agencies Appropriations for 2010*, Hearing (Washington, D.C.: US Government Printing Office, 2009).

US House of Representatives, Subcommittee on Energy and Environment of the Committee on Energy and Commerce, *The Role of Offsets in Climate Legislation*, Hearing (Washington, D.C.: US Government Printing Office, 2009).

US House of Representatives, Subcommittee on Energy of the Committee on Science, '*What Are the Administration's Priorities for Climate Change Technology?*', Hearing (Washington, D.C: US Government Printing Office, 2003).

——, *Department of Energy's Plan for Climate Change Technology Programs*, Hearing (Washington, DC: US Government Printing Office, 2006).

US National Academy of Sciences [NAS], *Energy and Climate: Studies in Geophysics* (Washington, D.C.: The National Academies Press, 1977).

——, *Policy Implications of Greenhouse Warming: Mitigation, Adaptation, and the Science Base* (Washington, D.C: The National Academies Press, 1992).

——, *Rising Above the Gathering Storm: Energizing and Employing America for a Brighter Economic Future* (Washington, DC: The National Academies Press, 2007).

——, 'Academy History', 2017a <http://www.nasonline.org/about-nas/history/> [accessed 19 October 2017].

——, 'Our Study Process. Ensuring Independent, Objective Advice', 2017b <http://www.nationalacademies.org/site_assets/groups/nasite/documents/webpage/na_069618.pdf> [accessed 19 October 2017].

——, *Negative Emissions Technologies and Reliable Sequestration: A Research Agenda* (Washington D.C.: The National Academies Press, 2019).

——, *Developing a Research Agenda and Research Governance Approaches for Climate Intervention Strategies That Reflect Sunlight to Cool Earth* (Washington, D.C.: The National Academies Press, 2020).

US National Research Council [NRC], *Weather and Climate Modification: Problems and Prospects; Summary and Recommendations; Final Report of the Panel on Weather and Climate Modification to the Committee on Atmospheric Sciences, National Academy of Sciences* (Washington, D.C.: The National Academies Press, 1965).

———, *Carbon Dioxide and Climate: A Scientific Assessment* (Washington, D.C.: The National Academies Press, 1979).
———, *Carbon Dioxide and Climate: A Second Assessment* (Washington, D.C: The National Academies Press, 1982).
———, *Changing Climate: Report for the Carbon Dioxide Assessment Committee* (Washington D.C.: The National Academies Press, 1983).
———, *A Review of the US Global Change Research Program's Draft Strategic Plan* (Washington D.C.: The National Academies Press, 2012).
———, *Climate Intervention: Carbon Dioxide Removal and Reliable Sequestration* (Washington, DC: The National Academies Press, 2015a).
———, *Climate Intervention: Reflecting Sunlight to Cool Earth* (Washington, D.C: The National Academies Press, 2015b).
US National Science Foundation [NSF], '65 FR 21795: Notice of the Availability of Draft Reports and Request for Comment' (Washington, D.C.: US Government Printing Office, 2000).
US President's Science Advisory Committee [PSAC], *Restoring the Quality of Our Environment* (Washington, D.C.: US Government Printing Office, 1965).
US Senate, 110th Congress, 'S.2307- Global Change Research Improvement Act of 2007', Version 22 May 2008 (Washington, D.C.: US Government Printing Office 2008).
US Senate, 111th Congress, 'S.2744 – Carbon Dioxide Capture Technology Act of 2009', Version 5 November 2009 (Washington D.C.: US Government Printing Office, 2009).
US Senate, 112th Congress, '*Carbon Dioxide Capture: Report to Accompany S.757*' (Washington, D.C.: US Government Printing Office, 2011)
———, 'S.757 – Carbon Dioxide Capture Technology Prize Act of 2011', Version 11 July 2011 (Washington D.C.: US Government Printing Office, 2011).
US Senate, 114th Congress, '*Energy and Water Development Appropriations Bill 2016: Report to Accompany H.R. 2028*' (Washington, D.C.: US Government Printing Office: 2015).
US Senate, 116th Congress, 'S.383 – Utilizing Significant Emissions with Innovative Technologies Act of 2019', Version 13 May 2019 (Washington D.C.: US Government Printing Office, 2019).
US Senate, Committee on Energy and Natural Resources, *Greenhouse Effect and Global Climate Change*, Hearing (Washington D.C.: US Government Printing Office, 1988).
US Senate, Committee on Environment and Public Works, *Global Climate Change*, Hearing (Washington, D.C.: US Government Printing Office, 1997).
———, *An Examination of the Impacts of Global Warming on the Chesapeake Bay*, Hearing (Washington, D.C.: US Government Printing Office, 2007).

——, *Climate Change: It's Happenging Now,* Hearing (Washington, D. C.: US Government Printing Office, 2013).

——, *Oversight Hearing: Examining EPA's Proposed Carbon Dioxide Emissions Rules from New, Modified, and Existing Power Plants,* Hearing (Washington, D.C.: US Government Printing Office, 2015).

US Senate, Committee on Energy and Natural Resources, *Future of Coal,* Hearing (Washington, D.C.: US Government Printing Office, 2007).

——, *Carbon Capture and Sequestration Legislation,* Hearing (Washington, D.C.: US Government Printing Office, 2011).

US Senate, Committee on Commerce, Science and Transportation, *Climate Services: Solutions from Commerce to Communities,* Hearing (Washington, D.C.: US Government Printing Office, 2009).

US Senate, Subcommittee on Energy and Water Development of the Committee on Appropriations, *Range of Innovative, Non-Geologic Applications for the Beneficial Reuse of Carbon Dioxide from Coal and Other Fossil Fuel Facilities,* Special Hearing (Washington, D.C.: US Government Printing Office, 2009).

US Senate, Subcommittee on Housing, Transportation, and Community Development of the Committee on Banking, Housing and Urban Affairs, *Public Transportation: A Core Climate Solution,* Hearing (Washington, D.C.: US Government Printing Office, 2009).

US Senate, Subcommittee on International Development and Foreign Assistance, Economic Affairs, and International Environmental Protection of the Committee on Foreign Relations, *Drought, Flooding and Refugees: Addressing the Impacts of Climate Change in the World's Most Vulnerable Nations,* Hearing (Washington, D.C.: US Government Printing Office, 2009).

US Senate, Subcommittee on Oceans, Atmosphere, Fisheries, and Coast Guard of the Committee on Commerce, Science, and Transportation, *Effects of Climate Change and Ocean Acidification on Living Marine Organisms,* Hearing (Washington, D.C.: US Government Printing Office, 2007).

Varmus, H., R. Klausner, E. Zerhouni, T. Acharya, A. S. Daar, and P. A. Singer, 'Grand Challenges in Global Health', *Science,* 302 (2003): 398–99.

Victor, D., 'Embed the Social Sciences in Climate Policy', *Nature,* 520 (2015): 27–29.

Ward, R., 'A Short Bibliography of United States Climatology', *Transactions of the American Climatological and Clinical Association,* 34 (1918).

Weart, S., *The Discovery of Global Warming* (Cambridge, Mass.: Harvard University Press, 2008).

Weaver, C. P., S. Mooney, D. Allen, N. Beller-Simms, T. Fish, A. E. Grambsch, and others, 'From Global Change Science to Action with Social Sciences', *Nature Climate Change,* 4.8 (2014): 656–59.

Weingart, P., 'Verwissenschaftlichung der Gesellschaft – Politisierung der Wissenschaft', *Zeitschrift Für Soziologie*, 12.3 (1983): 225–41.

———, *Die Stunde Der Wahrheit? Zum Verhältnis der Wissenschaft zu Politik, Wirtschaft und Medien in der Wissensgesellschaft* (Weilerswist: Velbrück Wissenschaft, 2001).

Weingart, P., A. Engels, and P. Pansegrau, *Von der Hypothese zur Katastrophe. Der anthropogene Klimawandel im Diskurs zwischen Wissenschaft, Politik und Massenmedien* (Wiesbaden: Leske + Budrich, 2008).

Weingast, B. R., 'Caught in the Middle: The President, Congress, and the Political-Bureaucratic System', in J. Aberbach and M. Peterson, eds, *The Executive Branch* (Oxford: Oxford University Press, 2005): pp. 312–43.

Wexler, H., 'Modifying Weather on a Large Scale', *Science*, 128.3331 (1958): 1059–63.

Williamson, P., D. W. R. Wallace, C. S. Law, P. W. Boyd, Y. Collos, P. Croot, and others, 'Ocean Fertilization for Geoengineering: A Review of Effectiveness, Environmental Impacts and Emerging Governance', *Process Safety and Environmental Protection*, 90.6 (2012): 475–88.

Wilson, E. O., *Naturalist*, (Washington, D.C: Island Press, 2006).

World Meteorological Organization [WMO], 'Yuri Antonievich Izrael', 2019 <https://public.wmo.int/en/about-us/awards/international-meteorological-organization-imo-prize/yuri-antonievich-izrael> [accessed 17 February 2021].

Woolgar, S., and D. Pawluch, 'Ontological Gerrymandering: The Anatomy of Social Problems Explanations', *Social Problems*, 32.3 (1985): 214–27.

Wyden, R., *History, Jurisdiction, and a Summary of Activities of the Committee on Energy and Natural Resources During the 112th Congress* (Washington, D.C.: US Government Printing Office, 2013)

Wynne, B., 'Seasick on the Third Wave? Subverting the Hegemony of Propositionalism', *Social Studies of Science*, 33/3 (2003): 401–417.

XPrize, '$100M Gigaton Scale Carbon Removal', *XPrize*, 2021 <https://www.xprize.org/prizes/elonmusk> [accessed 17 February 2021].

ACKNOWLEDGMENTS

This book has given me the great fortune to meet and work with a whole lot of wonderful people whose inspiration, support, and companionship I am deeply grateful for. These encounters and conversations each added further layers to this project, shaping it into what it is today.

This book started as a PhD thesis at the University of Bonn. I owe many thanks to my colleagues at the Forum Internationale Wissenschaft who have provided a wonderful environment for this research project to grow. I particularly want to thank my supervisors, Rudolf Stichweh and David Kaldewey, for granting me with the necessary trust, space, and resources. David Kaldewey and Daniela Russ continuously nourished my fascination with this project. Their questions, comments, and good sense have formed this work in many intangible ways. Our discussions in and out of the office sent me on a turbulent and somewhat odd journey through various strands of literature from systems theory, through science studies, and STS. David and Daniela read and commented on so many different bits and pieces, drafts and versions of this manuscript that I can only hope they like this final text and what I made of their invaluable commentary. The Stiftung Mercator has generously funded large parts of this research.

A set of expert interviews has provided valuable entry points and helped me to make sense of the wondrous world of climate engineering early on. I thank Jason Blackstock, Daniel Hayen, Joshua Horton, Hugh Hunt, Pete Irvine, Ben Kravitz, Jon Egil Kristjannson, Francesc Montserrat, David Morrow, Ted Parson, and Phil Renforth for their time and thoughtfulness. Our conversations brought some light into the dark and set me on track.

I was lucky enough that Jim Fleming agreed to participate in a small workshop which I hosted in Bonn during the early stages of this project. Since then, our exchanges have provided me with the necessary courage to embark on this historical-sociological journey. His scholarship and mentorship have pushed me down the rabbit hole of the 'usable history' of climate engineering. I thank him for paving the essential historical grounds for my sociological inquiry.

I had the pleasure to do much of the analysis and some of the writing of this book during a research stay in Boulder, Colorado. The Fulbright Commission provided a much-appreciated grant that made this field trip possible. Above all, I want to thank the scientists I had the pleasure of interviewing during my time in Boulder: Waleed Abdalati, John Barnes, Jim Butler, and Warren Washington. By taking the time and energy to attend to my questions and guiding me through your labs, you brought life to the oftentimes painfully dull policy documents, bills, and reports that I had been digging through for months. Max Boykoff and the entire team of the former Center for Science and Technology Policy Research at the University of Colorado, Boulder, have provided me with a wonderful research environment and good company. I especially want to thank Robin Moser, Brianne Eby, and Kevin Adams for making me feel home away from home.

A number of exchanges during the final stages of this project have helped me get this project through the finishing line. Kostis Chatziathanasiou has not only read and commented on every last bit of this text, but our conversations and his feedback have kept me going when it was the hardest. Ongoing conversations with Ina Möller have not only been of great pleasure but helped me tremendously in streamlining my argument and putting the finishing touches on my manuscript. Javier Lezaun has generously offered extensive feedback on several chapters of this manuscript. The attendees of two lunchtime seminars at Cambridge (UK) and Hamburg have provided their insights and critique on the final outline of this book.

I want to thank Mattering Press for doing a fantastic job. Two reviewers and especially my editor, Joe Deville, helped me turn a PhD thesis into a book that some people might actually want to read. Joe, our relentless back and forth is something I will never forget. Thank you for your time, energy, and thoughtfulness.

Finally, Kostis Chatziathanasiou, Daniela Russ, and Moritz Klenk have kept me afloat. Our conversations have been a source of life during the past months and years; you have ensured my happiness throughout all of it.

MATTERING PRESS TITLES

Environmental Alterities
EDITED BY CRISTÓBAL BONELLI AND ANTONIA WALFORD

With Microbes
EDITED BY CHARLOTTE BRIVES, MATTHÄUS REST AND SALLA SARIOLA

Energy Worlds in Experiment
EDITED BY JAMES MAGUIRE, LAURA WATTS AND BRITT ROSS WINTHEREIK

Boxes: A Field Guide
EDITED BY SUSANNE BAUER, MARTINA SCHLÜNDER AND MARIA RENTETZI

An Anthropology of Common Ground: Awkward Encounters in Heritage Work
NATHALIA SOFIE BRICHET

Ghost-Managed Medicine: Big Pharma's Invisible Hands
SERGIO SISMONDO

Inventing the Social
EDITED BY NOORTJE MARRES, MICHAEL GUGGENHEIM, ALEX WILKIE

Energy Babble
ANDY BOUCHER, BILL GAVER, TOBIE KERRIDGE, MIKE MICHAEL, LILIANA OVALLE, MATTHEW PLUMMER-FERNANDEZ AND ALEX WILKIE

The Ethnographic Case
EDITED BY EMILY YATES-DOERR AND CHRISTINE LABUSKI

On Curiosity: The Art of Market Seduction
FRANCK COCHOY

Practising Comparison: Logics, Relations, Collaborations
EDITED BY JOE DEVILLE, MICHAEL GUGGENHEIM AND ZUZANA HRDLIČKOVÁ

Modes of Knowing: Resources from the Baroque
EDITED BY JOHN LAW AND EVELYN RUPPERT

Imagining Classrooms: Stories of Children, Teaching and Ethnography
VICKI MACKNIGHT

www.ingramcontent.com/pod-product-compliance
Lightning Source LLC
Chambersburg PA
CBHW020834160426
43192CB00007B/639